Claude ROUQUETTE

THE PYRÉNÉES EUPROCT

Claude ROUQUETTE

THE PYRÉNÉES EUPROCT

Complex evolutionary processes

ScienciaScripts

Imprint

Any brand names and product names mentioned in this book are subject to trademark, brand or patent protection and are trademarks or registered trademarks of their respective holders. The use of brand names, product names, common names, trade names, product descriptions etc. even without a particular marking in this work is in no way to be construed to mean that such names may be regarded as unrestricted in respect of trademark and brand protection legislation and could thus be used by anyone.

Cover image: www.ingimage.com

This book is a translation from the original published under ISBN 978-620-3-42711-0.

Publisher:
Sciencia Scripts
is a trademark of
Dodo Books Indian Ocean Ltd. and OmniScriptum S.R.L publishing group

120 High Road, East Finchley, London, N2 9ED, United Kingdom
Str. Armeneasca 28/1, office 1, Chisinau MD-2012, Republic of Moldova, Europe
Printed at: see last page
ISBN: 978-620-7-71814-6

Contents

The naturalist suite on the complex processes of biological evolution and the transformations of civilisation consists of four chronological volumes:

Volume I: **Les fougeres noires. (To be published)**

Volume II: **L'Euprocte des Pyrenees.** (*Editions Universitaires Europeennes, 2021* **).**

Volume III: **Le Castor des Cevennes** (*Editions Universitaires Europeennes, 2021*).

Volume IV: **Indri Indri, a journey to the origins of humanity (To be published)**

Followed by three essays:

- **Low biological energies.** (*Editions Universitaires Europeennes, 2021)*
- **L'Humanite desarticulee (To be published).**
- **La septieme corde (To be published).**

Other works by the author on the history of the navy and the Cevennes, an introduction to scientific exploration of the seas, oceans and continents.

- **Neptune's college.** (*Les Presse du Midi).*
- **Memoires de mer, cevenoles. (To be published).**
- **Besseges, in times of plenty, 1825-2021, decline and renewal of a Cevennes town** (*La Fenestrelle*).

This logbook is a tribute to our friend

Jean Gagnepain (1961-2010)

Vice-president of the *"Pour Darwin"* association and Director of the gorges du Verdon prehistory museum in Quinson, he has generously supported the activities of the Charles Darwin Institute International.

Acknowledgements

To Professor Patrick Tort, Director of the Charles Darwin International Institute, with all my gratitude for his kindness and scientific rigour.

Professor Jean-Claude Beetschen, for all his work.

At the Cëvennes National Park documentation and archive centre in Gënolhac.

At the musëe de Prehistoire des gorges du Verdon.

To my friends, Suzette and Henri Blandina, to whom we owe the Ethnobotanical Gardens of La Gardie in Rousson (Gard, Occitanie), where the terrestrial salamanders live.

The musëe de Prehistoire de Rousson, for its collection of amphibian fossils.

To the wildlife photographer Jean Faisse and Michele, for their long observations of amphibians and reptiles on Mont Lozere, and other horned animals.

To my comrades, sailors, naturalist guide-confërenciers, Sylvain Mahuzier and Jean-Pierre Sylvestre.

To Capitaine de vaisseau (h) Jean-Claude Richard, former head of the ëlectromëcaniciens de sëcuritë course, who prëfacë the essay on low biological ënergies, complementary to this tome.

To Admiral (2s) Christophe Prazuck, Director of the Institut de l'Ocëan, Sorbonne Universite for his encouragement to pursue scientific exploration of the seas, oceans and continents.

A la Dame, a l'ancre.

from
Jean-Pierre Sylvestre

An international specialist in marine mammals,
which he has studied in all seas and oceans.
Antarctic conference guide.
Author of numerous articles and books on cetaceans.

The marine historian and naturalist Claude Rouquette has asked me to write the preface to his impressive research on the Pyrenean skink. I am deeply touched by this long-standing friend's request and his confidence in me. I agreed to take it on as a specialist in marine mammals who has no real knowledge of present-day amphibians. In memory of our friendly discussions on Natural Sciences during my military service in the French Navy, I wanted to write these few preliminary pages by commenting on the phylogëny of amphibians, that which links certain Fishes to the Tëtrapods. Since I was very young, I have been fascinated by the Crelacanth, which I have been lucky enough to study in numerous museums around the world, in South Africa, Japan, Germany and France. My encounter with scientists working on this 'primitive fish' taught me that there are only two living species, namely :
-    The Indian Ocean and Mozambique Channel Crelacanth, ***Latimeria*** *chalumnae.*
-    Crelacanthe de l'lndon'sie, ***Latimeria*** *menadoensis.*

Often referred to as a "living fossil", these two present-day species represent a recent branch of this family, but this very special fish has nevertheless preserved in its anatomy significant traces of an evolutionary pathway between pulmonid fish and the first terrestrial vertebrates or Tëtrapods, the origin of the Amphibians. For example, the pectoral and pelvic fin bones of the Crelacanth are distinguished by ancestral bones, homologous to our front (arms) and rear (legs) limbs.

Scientists in the second half of the 20th century saw this "lobe-finned fish" as the ancestor of terrestrial vertebrates, and therefore of amphibians, then of reptiles, as well as mammals, going back from primates to Homo sapiens!

As mëthodes of phylogënëtic classifications have improved, they have сотдё the taxonomic arrangement of fishes by sëparating species into several well spëcific clades :
-    ***Osteichthyes***, including bony fish with swim bladders, such as the Sardine (***Sardina*** *pilchardus*) and the Mërou of the vast family ***Serranidae.***
-    The ***Chondrichthyes***, which include the cartilaginous Fishes including, the subclass ***Elasmobranchii*** of Sharks and Rays, and ***Holocephali*** of Chimëres, without swim bladders.
-    ***Agnatha***, or jawless fish such as the Lamprey.
-    The ***Sarcopterygians***, formerly called Crossoptërygians, from which dërivë derive the ligiwes of the ***Ccelacanthiformes*** with lobëe fins and the Dipneustes, of the super-order ***Ceratodontimorpha***, in which the swim bladder is equipped with functional cHilvroles for breathing in the open air.

The cmraeteres acquired by the Crelacanthus suggested evolutionary patterns during the transition between the aquatic vertëbrës and the first quadruped terrestrial vertëbrës.

In Quebec, I'm lucky enough to live not far from a fossil site of prime importance for pa^ontology research, namely Miguasha in Gaspësie. At this site, researchers have discovered one of the most ëtudied fossil species of fish in order to understand its adaptability when it emerged from the water during the Dëvonian, a warm, dry and very selective period 380 million years ago. It is thought to be a Sarcoptërygian, a cousin of the Crelacanth: ***Eusthenopteron*** *foordi Whiteaves, 1881*, baptized the "Prince of Miguasha" by Q^becquois scientists, who considered it to be the closest form of the first vertebrës prëaptës to terrestrial life, and therefore close to the Amphibians. On the same site, between 1950 and 1981, Canadian pa^ontologists unearthed two very fragmentary pieces of another aquatic species that looked more like an Amphibian than a Fish, named ***Elpistostege*** *watsoni,* with pectoral fins comparable to fingers. In 2010, they discovered a whole spëcimen that was sufficiently well preserved to place it in the line of descent between fish and amphibians.

Other spëcimens will be found in the Dëvonian gëological layers in the Northern Hemisphere, notably in the Canadian Arctic with ***Tiktaalik*** *roseae*, which lived in shallow, oxygen-poor waters and could move out of the water supported by its strong pectoral fins and breathe air using its highly vascularised swim bladders. In Russia and Latvia, ***Panderichthys*** *rhombolepis* and *stolbovi* had primitive lungs and moved along the ground using the combined movements of their lower limbs and undulation. These three species classified as Sarcoptërygians raised a question about their classification: were they lobe-finned fish or primitive amphibians?

The pa^ontologists carried out in-depth anatomical studies on the pectoral fins of ***Eusthenopteron, Elpistostege, Pandericthys*** and ***Tiktaalik***, which confirmed their evolutionary position between the Pulmo^s Fishes and the first terrestrial vertebrës Tëtrapods, in the direction of the Amphibians. Their research confirmed the description of an ancestral form of the humërus, radius, ulna and distal ëlements in the anterior limbs. Sufficiently valid clues to presage a missing link between Pisces and Amphibians. These aquatic vertebrates seemed to have made the transition by having the characteristics to breathe air and move from the aquatic environment to dry land.

This emergence from the water was confirmed by the discovery in the Northern Hemisphere of amphibian fossils found in Upper Devonian layers dating back between 385 and 259 million years, the best known of which are ***Ichthyostega*** and ***Acanthostega*** from Greenland.

My interest in the Crelacanthus enabled me to acquire a greater knowledge of fossil amphibians than of the present-day forms studied by the naturalist Claude Rouquette as part of his research into complex evolutionary processes.

In Quebec, I had the opportunity to work on a Proteid: the Spotted Necture, ***Necturus*** *maculosus*, a Urodele that retains its gills throughout its life, as is the case with the Mexican Axolotl and the Slovenian Proteid.

In Japan and China, I studied two species of giant salamander, ***Andrias*** *japonicus* and ***Andrias*** *davidianus*, and during several visits to the Hiroshima region, I observed and photographed these giant salamanders in the streams to the north of the city. I would like to thank my friends at Asa Zoo who are working to protect this species of Japanese fauna.

In keeping with my friend's research, I would like to comment on the similarities and differences between these giant salamanders, isolated in a few streams on the Japanese island, and the Pyrenean Euproctus. These species of Urodeles live in mountain streams with cold, oxygenated water; their mating is nocturnal, and they hide during the day, away from the light. In comparison, the small Euproctus *Calotriton asper asper* manages to slip under rocks to shelter from the sun's rays, while the large Japanese Salamander (between 80 cm and 1.40 m) seeks out hiding places underwater vegetation or in rock crevices. In the evening, as soon as night falls, the Japanese salamander emerges from its aquatic hiding place and walks over the gravel in search of prey.

While the giant salamanders of Japan and China are the largest species of Lissamphibians living on earth, the fossil forms of the Paleozoi'que and Mesozoi'que were much longer. In North America, scientists describe two fossilised amphibians 2 metres long, one dating from the Devonian, *Hynerpeton*, and the other from the Permian, *Eryops megacephalus*. In Europe, the Germans mention a 2-5 metre long amphibian, *Mastodonsaurus*, found in the Upper Triassic. In Australia, researchers have discovered a gigantic 10-metre-long salamander, *Koolasuchus cleelandi*, in Cretaceous strata.

During my geological travels around the world, I had the opportunity to spend several days at one of the richest paleontological deposits from the early Permian, between 290 and 260 million years ago. It is located at Rotliegen in the Saarland, a region of Germany wedged between southern Luxembourg and France. In the many layers of this site, researchers have discovered freshwater aquatic species including Crelacanths, various bony fish and above all primitive sharks, some of which were over 3 metres long.

On land, they found insects, a large dragonfly and, above all, amphibians ranging in length from 20 centimetres to over a metre. Some of these amphibians resembled large salamanders with crocodile skulls. Thus, the flattened, triangular skulls of the amphibians *Sclerocephalus* and *Archegosaurus decheni* (which is thought to have been more fish-like) not only resembled that of the alligator but also resembled the Quebec *Elpistostege*, following a process of evolutionary convergence. German paleontologists described small salamanders, such as *Apateon* (12 cm), which was totally aquatic and covered in scales. These salamanders resembled the Canadian necture and other fossil species morphologically similar to the Pyrenean Euprocte, including *Micromelerpeton* (300 Ma), which could regenerate its limbs, and *Melanerpeton*.

The study of amphibian fossils, paleobatrochology, began in German-speaking Switzerland. In 1725, some men brought the Swiss naturalist Jacob Scheuchzer (1675-1733) a limestone slab discovered in a German quarry near the town of Ohningen, on Lake Constance. The slab showed a skeleton almost a metre long, barely emerging from the rocky mass. The naturalist from Zurich identified a skull and vertebral column. On examining the remains, Scheuchzer suddenly became convinced that this was a man who had died at the time of the biblical flood. He published a detailed description of it in his "*Litographia helvitica*" of 1726 and named it *Homo diluvii testis, a* man who had witnessed the flood and who, according to the precepts of Natural Theology laid down in the writings of the Old Testament, intended to set the upper limit to the creation of species. It was not until 1812, with a description by the famous French scientist Georges Cuvier (1769-1832), that the name of this Miocene salamander, *Andrias scheuchzeri*, which he cautiously described as Scheuchzer's human image, was confirmed... Since 1802, this specimen has been on display in the Teylers Museum in Haarleem, the Netherlands.

By studying the manuscript of the Euprocte des Pyrenees, I have completed my knowledge of the ways in which amphibians make the transition between aquatic and terrestrial environments, which I have just briefly outlined by summarising a few features of their paleontology. The adaptability of respiratory processes in Urodeles merits further study, and I was impressed by Claude Rouquette's exploratory research into complex processes in the biology of revolution and the transformations of civilisation. His investigation of the units of levels of integration of living organisms, together with his many questions, provide us with information on the complementary nature of cutaneous, oral-pharyngeal and pulmonary respiration in anurans and urodeles. To explain this, he has chosen to venture into a prospective description of the co-ordination of cellular activity, examining its causes and consequences on the ontogeny, metamorphosis and regeneration capacity of the Pyrenean Euproctus, and concluding with a discussion of their nascent sociability.

We met in 1981, when Master Chief Claude Rouquette was one of the instructors in the Nuclear, Bacteriological and Chemical (NBC) Defence department at the Querqueville Naval Training Centre, then the French Navy's security school, where I would be posted during my military service, along with Seaman Sylvain Mahuzier.[1].

We had a passion for Natural Sciences and Cryptozoology, and we were lucky, because this department was made up of naval officers, petty officers, scientific assistants, doctors and pharmacists who provided NBC risk prevention training on board the combat ships. Among them was Claude Rouquette, with whom Sylvain and I had very long discussions on zoology and the theory of the evolution of species, which he studied during his round-the-world voyages. This is how the three of us came to form this strong friendship, the friendship between the crews of the Fleet and the 'on-board' naturalists, a solidarity that is given concrete expression in this preface.

Jean-Pierre Sylvestre

ORCA, l'Isle-Verte, Quebec, Canada
(October 2021)

---

1 Grandson of explorer Albert Mahuzier, Sylvain is a biology teacher, consultant naturalist, lecturer and author. In 2016, we co-wrote "Cap sur le grand continent blanc", published by QUAE. Sylvain prefaced the first volume of Claude Rouquette's naturalist suite, 'Le castor des Cevennes', published by Editions Universitaires Européennes, as an introduction to complex evolutionary processes.

Foreword by a naturalist

Sailing in the service of the Royale[2] in the Mediterranean, the Persian Gulf and the Indian Ocean, gave me a few respites between two pëtroliëres crises, the time for a careening and a short leave. Even during these brief përiods of respite, interrupted by untimely recalls, the political and military ëvënements caught up with us on the paths of the Hautes-Pyrenees that we travelled from 1983 onwards, close to the Neouvielle reserve, in the Moudang valley, above Saint-Lary-Soulan. In 2009, this experience was repeated as we began to develop our method for working on complex evolutionary processes.

Many questions have arisen in the course of writing this volume of the naturalist sequel, and you will see that we are asking more questions than we are answering in increasingly specialised areas of science that are beyond our grasp. In order to write more substantial reports than these few chapters, taking our inspiration from Charles Darwin's successful journey, we have proposed a joint project with the Charles Darwin Institute International, under the direction of Professor Patrick Tort. In the spirit of Charles Darwin's voyage, we have suggested to the Institutions that they set up a fleet of sailboats equipped with on-board laboratories, travelling vectors and autonomous modules dedicated to multidisciplinary scientific research. We recall the general objectives of long-term explorations in *"Biological Evolution and Transformations of Civilisation"* for missions conducted simultaneously at sea and on land, without interruption.

"The Charles Darwin Institute International has set itself the goal of continuing the work of the great naturalist on the seas, oceans and continents, to measure the revolution of ecosystems and biogeographical zones. The aim is to study the impact of their transformations on human life, which is integrated into them today, and to draw the ecological, cultural and social consequences in order to answer vital questions about the future of Civilisation, the evolution of Humanity, living species and their environments".

Our first naturalistic stop, at the end of a mountain path in the Moudang valley, was to be used to try out the exploratory method for complex evolutionary processes.

Stopovers in the Moudang valley

During our exploratory escapades in the I fiutes-Pvrenees, we followed in the footsteps of Charles Darwin, who travelled through the Andes before reaching the Galapagos Islands, and in the footsteps of other naturalists who inspired our approach. The biologist Claude Dendaletche, writing in his travel diaries, opened the way for us to discover the ecosystems of the Pyrenees. As for the scientists, Raymond Despax, Monique Clergue-Gazeau, Jean-Claude Beetschen and Francois Gasser, their in-depth research into Urodeles encouraged us to humbly follow in their footsteps, to discover and reveal certain aspects of the Euproct, a species of amphibian endemic to the Pyrenees, that are still too little known. On the border between France and Spain, these mountains are not too high to make access to the highest peaks practicable. The Vallee d'Aure offered a wealth of biodiversity accessible at different altitudes, making it an ideal location for long hikes and fascinating observations for a novice naturalist. It was in the course of the Neste in the Moudang valley, a torrent of limpid waters where tumultuous waterfalls pour down and ferruginous springs are born, that we met...

The Pyrenean skink, *Euproctus or Calotriton* asper asper *(Duges, 1852)*

Our base camp was at the Pont de Moudang campsite, a hamlet above Saint-Lary-Soulan, the very welcoming spa and ski resort in the Hautes-Pyrenees. The aim of our stay was twofold: to carry out historical research into the route taken by shipbuilding timber harvested in the forests of the Pyrenees during the era of the sailing navy, and to begin developing a method of observation on a site with features conducive to the practical study of complex evolutionary processes.

In 1991, my first meeting with Professor Patrick Tort at the conference he had organised at the College de France and the Sorbonne on the theme of '*Darwinism and Society*' was decisive. In his opening lecture, he explained the foundations of Darwinian anthropology, emphasising the fundamental notion of the reversive effect of revolution, a major concept that he had set out in 1983. These sessions took place in front of an audience that was always curious, looking for explanations on the considerable extensions of the famous scientific communication expressed by Charles Darwin, on the origin of species by natural selection, in conjunction with the assertions of Alfred Russell Wallace. The speakers went on to explain the effects of the social aberrations of Darwinism caused by ignorance of the complete work of the learned naturalist, which is still too little known and often misinterpreted. Dominating the audience with his ever vigilant listening, the wise Theodore Monod paid close attention to the words of Professor Patrick Tort, whose innovative and original discourse was still difficult for a sailor and amateur naturalist to grasp.

I was still asking myself a lot of questions about the biological evolution I was discovering and about civilisation, which was always in crisis, and with which I was frequently confronted during incessant overseas operations centred on the Mediterranean, a conflict-ridden maritime zone that extended from the Near to the Middle East. These questions prompted me to pursue in-depth studies on the theory of revolution, taking into account this singular paradigm that Professor Tort introduced and would tirelessly explain through impressive publications such as the '*Dictionnaire du darwinisme et de revolution*' (puf, 1996) and the competing re-edition of Charles Darwin's work, whose erudite prefaces he rigorously rewrote. Other works by Patrick Tort caught my attention, in particular '*La Pensee hierarchique et I'Evolution*' (Aubier, 1983), a decisive text on his method of analysing discursive complexes, a major theme in his work, which he developed at length in '*Qu'est-ce que le materialisme*' (Belin, 2016), a consequent essay that I would read in homeopathic doses to reinforce my research into complex evolutionary processes.

The analysis of discursive complexes is a method that rightly defies historical situations as they relate to the sciences, which are fundamentally creative, but whose discourse is abusively impregnated by reiterative ideologies in an inexorable reshuffle that skilfully manipulates scientific invention. In order to achieve their ends, and no less skilfully in the case of the most obscure beliefs, to neutralise them.

With this warning in mind, the linguist and historian of science challenged me on the true dimension of the theory of biological revolution, extended to that of Civilisation. A Civilisation in permanent crisis, whose most perverse effects we were to measure in concrete terms for many long years to come.

Those of the incessant warlike conflicts in the Middle East, the war between Israel and its close neighbours (the Yom Kippur War, 1973), which led to the annihilation of Lebanon, then the oil war which followed from Iraq to Libya, Iran, Syria and Afghanistan, without respite...

Those of technological and chemical disasters (Bophal, 1984) and nuclear disasters (Chernobyl, 1986).

Those of the socio-economic decay of the Soviet regime contrasted with the ultra-liberal exuberance of the United States.

Subjected to the insidious effects of globalisation, the States of Europe and France will not be spared, as the challenges of galloping modernity and out-of-date finance increase social fractures, generating austerity and growing precarity. Jobs became scarce, opening the way to delinquency and crime, and the rise of Islamic terrorism spread insidiously to strike cruelly at populations in order to destabilise countries suffering the effects of uncontrollable migratory flows. Events that encouraged extremist tendencies on all sides, triggering cruel and endless conflicts...

From 2020 onwards, the global health and climate crises will shake up the governments and peoples of every country...

In the service of the nation and as a citizen of the world, I had become an attentive witness, at the time neutralised by the duty of reserve. I lived through these conflicts, both abroad and in France, on board the combat ships of the French Navy. The naval forces were engaged alongside our brothers in arms of the Air Force and the Army, and we would support each other in the worst moments, during the attempts to find diplomatic and armed solutions to these tragic problems of civilisation. These episodes will be interspersed with short reparative sëdays in the peaceful Pvrenean mountains, interrupted by a tëlëgramme of recall. International circumstances and the demands of government service at sea led us into less peaceful terrain than our periodic naturalist explorations along the paths of the Pyrenees, in the Moudang valley.

1. *Hautes-Pyrenees, the Moudang valley.*

Our armed forces were embroiled in a quasi-permanent war, from the disintegrated Lebanon to the explosive Persian Gulf, situations in which the French naval air force was constantly involved. The aeronautical group was made up of an aircraft carrier, and alternately, I embarked on the *'Foch'* or the *'Clemenceau'*, on which I was assigned as Major of the security speciality. This powerful ship, *the 'Tiger of the Seas'*, was surrounded by its escort, the frigate *'Suffren'* and a corvette, supported by a supply ship. This air and sea force was protected by its own aircraft and probably by a few discreet submarines in the water. Under the command of an admiral, this squadron's mission was to defend our interests abroad by supporting France's diplomacy while ensuring the free flow of oil supplies. A natural resource and a fossil fuel wasted on the highways of consumerism and carefree leisure, the profits from which are destined to provide for the reckless profits of multinationals and their wealthy shareholders who, judging by the size of their yachts, are not yet satisfied with their insatiable desires.

At each stopover, I perceived the multiple consequences of limitless progress, on social life, on the environment and on the revolution of species, I noted the potential excesses of the developed countries and noticed the shortcomings of those considered to be developing, paradoxical situations, with discrepancies that seriously alter civilisation. I observed the misery of the peoples caught up in the endless armed conflicts affecting the Mediterranean basin, an explosive cauldron of an ever-cold war in which the powerful European naval forces (France, England, Italy, Spain, Greece) and the fleets of the United States were concentrated, Greece) and the fleets of the United States, confronted by those of the still Soviet Union, which harassed us relentlessly, from the high seas to the shallow anchorages, in the interminable wait for an engagement, delegated to a few militias bribed into fratricidal combat.

The historical observation that the human race is capable of aggression and destruction merited in-depth reflection, and a great deal of reading in ethology (Lorentz, 1963) and behavioural biology (Laborit, 1973) punctuated the long days at sea. After the watch, the aviations, the safety drills and the recalls to battle stations, it was sometimes a quiet hour of reading in the narrow cabin of the aircraft carrier. A real resource for the mind, freeing it to escape momentarily from the many constraints of maritime and military life, in order to better appreciate the unforgettable discoveries made during the short refuelling stops. As part of my internal promotion, I also had to devote some time to preparing for the competitive examination to become a naval officer specialising in safety and the environment, by taking correspondence courses, and thus become interested in the workings of national, European and international institutions, as well as the history of the Navy, France and the world in all its states... At the time, I was studying the revolution of the tiger (**Panthera** *tigris*), which was already on the verge of extinction.[3] At the time, I was studying the very particular revolution of tigers (Panthera tigris), already on the verge of extinction, as well as animal representation, an art form that was very abundant around the Mediterranean, and more particularly in Greece, certainly seduced by Mycenae and its archaeological treasures dominated by the presence of lions (**Panthera** *leo*). The research I was carrying out was based on the original representation of the evolutionary history of the Felidae that the paleontologist Georges Gaylor Simpsom had judiciously introduced into some very elaborate phylogenetic tree diagrams, which I found very intriguing. For this line of the family Felidae, he had traced three evolutionary trajectories, those of the Viverrids, Felids and Machai'rodontes, which effectively represented the progression of their morphological variations in relation to natural selective constraints that he had surprisingly stratified into an explicit diagram of their evolution and extinction. I completed this summary study of the tiger revolution by exchanging letters with the American paleontologist Leonard Radinsky (1937-1985), who had devoted his research to the functional analysis of the morphology of sabre-toothed tigers, a critical hypertelia caused by the excessive elongation of their canines, which he believed contributed to their extinction by limiting access to certain prey. According to his hypotheses, their skin became too thick to be pierced by these long teeth, due to the combined effect of the determining ratio of genetic variations and natural selection on this vital feature of its dentition, which had become hypertrophied. To come back to the notion of aggression, the aggressive instinct of tigers merited further study, and while taking advantage of a long docking trip to Toulon, I had the opportunity to observe and record tiger behaviour at the Faron breeding and training centre.

Following the advice of his owner and friend Roger de Souza, who was used to the aggressive behaviour he provoked during training sessions. This skilful wild animal trainer knew how to trigger and use the tiger's anger: when it approached the cage from a distance that disturbed it, the tiger would suddenly stand up on its hind legs, fangs bared and claws out, ready to destroy its prey with frightening roars and howls. I took note of the phases in which this powerful feline's aggression was triggered, attentive to the slightest movement of the other animals and visitors in the animal park.

This aggressiveness ranged from simple dissuasion to terrorising brutality, which manifested itself for several reasons that I had to understand in natural environments... Later, while visiting southern Nepal, during a stay in the Terai at Lumbini, the birthplace of the historical Buddha who devoted his life to finding and teaching the quest for inner peace. According to his doctrine, the mastery of instincts and desires is based on meditative practices in order to achieve a calming of the senses and thoughts. We were on the border with India, close to the Royal Chitwan national park, and every morning in the local newspapers I read about tiger attacks on locals, guides and tourists. These murderous attacks are one of the major problems for wildlife conservation, problems that I would find again with the bears of the Himalayas, from Nepal to Bhutan, as with those exterminated

---

3 Panthera tigris is on the brink of extinction, with around 3,800 remaining in Asia.

and reintroduced, not without difficulty, in the Pyrenees. On the strength of these experiences, I felt it was important to translate the acquisition of book knowledge on the theory of revolution into practical studies, in order to develop a method of observing and recording data that could one day be used to reflect on and formulate proposals for the complex processes of biological evolution and the transformations of civilisation.

After successfully writing an essay on Konrad Lorentz's *"L'Homme dans le fleuve du vivant"*, Homme qu'il pressentait deja en peril, (Flammarion, 1961) in an internal competitive examination, I was admitted to the Ecole Militaire de la Flotte, the gateway to the Ecole Navale. Having become a midshipman, the training period for naval officers on the Crozon peninsula came to an end, and I boarded the aircraft carrier *'Clemenceau'* as a security assistant. A year after this last embarkation, I was promoted to the rank of ensign and was appointed first officer and then company commander in the prestigious Bataillon des Marins-Pompiers de Marseille. This maritime posting allowed me, after having spent a long time travelling the seas and oceans, during the endless conflicts of the Cold War, often burning in the great theatre of human tragedy, to see society from the inside. As a result, I was able to visit Marseille's libraries and museum more regularly, as well as the bookshop in the old Galeres arsenal, and to listen to lectures and take part in cultural events in the charming city of Phoceenne. It was a great experience of command, working alongside the courageous and dedicated sailors and firefighters. It was certainly challenging, but it was enriched by the many contacts with the people of Marseilles, who are confronted with the challenges they face. From this observatory of everyday sociology and through my experiences in the field, in working-class and more affluent neighbourhoods, human distress was revealed, I noted the deep causes of the collapse of society, after having experienced those of the United Nations, facts that I comment on in the essay on *"L'Питание desarticulee"*...

These exciting and instructive activities had so оссиpё me during the many years spent in the service of the Royal... Reached by the age limit, I was increasingly concerned by the problems of Humanity and curious about the complexity of its biological evolution towards Civilisation. A civilisation constantly undermined by the incessant conflicts around the world and the social and economic dysfunction revealed to me by my many trips around the world and my last shore-based assignments. It was time to leave the noble maritime institution and devote myself full-time to naturalist observations and research into complex evolutionary processes. The founding of the Charles Darwin International Institute in Puycelsi, in the Tarn region of France, directed by Professor Patrick Tort, was to provide me with the opportunity to deepen my knowledge of the great naturalist, which was far too succinct, after an in-depth study of the History of the Navy and of the voyages of scientific exploration that preceded and followed that of Charles Darwin.

2. *Library, Charles Darwin Institute International.*

In the company of Patrick Tort, we will be conducting fascinating research into the three voyages of the H.M.S. '*Beagle*', in search of the beginnings of a theory that the young naturalist developed on the second voyage of this small three-masted barque, which sailed from South America (1825-1836) to Australia (1837-1843), ending up at the bottom of an arsenal (1872) after a circumnavigation of the globe. This study, carried out over more than three years, inspired us to design and develop projects to promote long-term scientific exploration, and encouraged me to pursue research into the complex processes of biological evolution and the transformations of civilisation. Between my travels and my retirement in Cevennes, I also had some fruitful encounters within the '*Pour Darwin*' association, whose members from very different backgrounds support the activities of the I.C.D.I. alongside its indefatigable director.

This was the continuation of a friendly complicity and a sustained dialogue with Patrick Tort, who led us to contribute to the writing of the long preface to Charles Darwin's "*Logbook* [4]", in particular to describe in detail life on board the famous little sailing ship. I highlighted the system of relationships, typical of people at sea, that had been established between Charles Darwin, his assistant, the officers and sailors of the H.M.S. '*Beagle*'. This stimulating context, alternating between navigation, hydrographic surveys of the coasts and explorations on land, was in my opinion intellectually decisive for his discovery.

At the time, I was writing the manuscript of the College de Neptune, the history of one of the first '*Naval Schools*' set up in Ales in the Cevennes before the Revolution, as well as researching documents on the officers of the H.MS. "*Beagle*'s officers, and in particular information about her commanding officer. By 1860, having been promoted to admiral, Robert FitzRoy had distanced himself from Charles Darwin, who was irreconcilable with this rigorous follower of biblical texts. However, paradoxically to this fundamentalist attitude exacerbated by the rigid exercise of command, he had become a brilliant meteorologist and would be the initiator of the first weather forecasts (1861) as well as the inventor of a code using conical signals to warn ships of storms. In this capacity, he maintained correspondence with officers of the French navy and with the Academie des

---

4 Charles Darwin, *Le journal de bord du voyage du Beagle (1831-1836)*, preceded by Patrick Tort with the collaboration of Claude Rouquette, editions Slatkine et Champion, 2011.

Sciences, letters that we had to find to support our arguments in the preface to Charles Darwin's logbook... The detailed study of his letters revealed to me the young scientist-naturalist's method of exploration, which he carried out at sea and on land, filling notebooks and drawing on the small ship's library. Inspired by his approach, after some fifteen years of development and field observations, our method of analysing complex evolutionary processes was beginning to bear fruit. We thought it would be a good idea to consolidate our field research within the attractive framework of the naturalist tradition, without abandoning the scientific rigour demanded by my approach, which took advantage of advances in scientific disciplines, and to publish it in the form of a naturalist sequel, "*Biological Evolution and the Transformations of Civilisation*", in four volumes, accompanied by three essays, including a supplementary essay on "*Low Biological Energies*" to explain passages in the present volume. This will be followed by an essay on '*Humanity Disarticulated*', an essential recapitulation of my long investigations, before writing a final synthesis on complex evolutionary processes, a sort of '*Seventh String*' to declare the notes of a necessarily unfinished score.

The volume devoted to the Cevennes beaver enabled us to work on the modelling and assimilation of data from a probable case of speciation by proposing the fundamental notion of '***adaptability***' resulting from the highly non-linear nature of complex evolutionary processes, and above all, to dare to integrate the concept of the reversal effect with the usual caution. A powerful agent of revolution in the human species, as a singular indicator of the trend reversal, from social instincts towards Civilisation, represented to this day in a theoretical and didactic way by the Mobius ring. A fundamental torsion derived from topology, from which it was necessary to draw inspiration in order to design a more operational model designed to explore and compare the sociable instincts of the species studied, from amphibians to rodents to primates, in order to understand civilised humanity.

At this point in our research on complex evolutionary processes, described in Appendix 1, we had a stochastic cursor to probe the biological evolution of a species in order to make cross-sections and observe with a magnifying glass the zones of phylogënëtic divergence of mammals, and in this case of amphibians...

Finally, I was going to use the powerful reversal effect opërator with its rërivëes rëvëlatrices which tend, through the singular orientation of the Primates' social instincts, to free themselves from natural sëlection by crossing that Rubicon, so original and exclusive, with which the Hominins accede to civilisation. The aim of this experiment was to analyse the clues leading to the beginnings of social instincts, which have become culturally effective and exceptional to the Hominin branch, but which preexist in the animal world at different levels of expression of their individual and collective behaviour. Levels of cooperation and sociability that had to be identified and estimated for each species studied: from Amphibians to Mammals, from Rodents to Primates, going back to the origins of Humanity, in the company of the Lemurians of Madagascar.

The all-too-rare observations made of the Pyrenean skink provided an opportunity to ask further questions about the biological variability of this species in the face of the many natural and anthropogenic selective constraints that apply at different altitudes. It was also an opportunity to present the first terms of the foundations of an apparently sociable behaviour by getting out of the traps set by sociobiology, which is obsessed with continuity, the influence of genes on individuals and their presumed hold on the population of a species.

Firstly, it was necessary to identify the initial conditions for the revolution of the Euprote, going back to the origins of a species with an ancient lineage that undergoes a metamorphosis in the course of its ontogeny, depending on the environmental conditions at altitude, and which also has the ability to regenerate its organs, which are incidentally mutilated.

In order to use the method for analysing, modelling and simulating complex evolutionary processes, I had to record in the field a number of parameters to be included in the Darwinian ratio of random variations to selective constraints in order to establish updated coefficients of adaptability. While observations of **Castor** *fiber* continued in the Cevennes, in 2009 we returned to the Pyrenees massif to carry out further surveys of Euproctes.

The last round-the-world journeys, finally accompanied by my wife, took us from the Himalayas to Madagascar, to finally accomplish, in the company of the Lemurians, this essential return to the origins of Humanity, in order to try to answer the vital questions raised about the future of civilisation, species and their environment. This is our aim in the naturalist suite and in our essays on complex evolutionary processes, particularly in this volume devoted to the Pyrenean puffin.

In Chapter I, we must focus on the orogeny of the Pyrenees massif. Its formation provides us with information about its current geology and the various changes that have affected the life of living organisms, the flora and fauna, and the men and women who populate the Pyrenean mountains. The biogeography and history of this territory will be highly dependent on these geological formations, which introduced a natural frontier between Spain and France.

This geological base and the climate have shaped the high mountain landscape, which is the source of the traditions and customs of each valley and of the socio-economic life of the mountain dwellers who have shaped the Pyrenean landscapes of today. However, climate and ecological changes are accelerating, as human activities interact surreptitiously, including in the wetlands where the Euproctes live discreetly.

In Chapter II, we describe the Moudang valley, a majestic site surrounded by peaks, still preserved from invasive human activity, the impact of which can be seen as we travel up from the Aure valley towards the Neouvielle nature reserve. The nearby Granges du Moudang hunting and grazing reserve, bordered by the River Neste, is an ideal place to study the revolution in species living in contact with a pastoral civilisation that has left its mark on the landscape (Dendaletche, 1982). As we wandered along the stony paths beside the Neste, pausing at the edge of a stream, we discovered the Euproctes. An aquatic salamander with a wide range of abilities, capable of developing and metamorphosing in response to temperature and altitude, and of regenerating an organ in the harsh environment of the deep valleys of the high peaks surrounding the Moudang barns.

In Chapter III, we draw up an identity card for the Pyrenean snake, **Calotriton** *asper asper,* which we refer to as the snake in our presentation. This species, which is very common in small streams, does not always attract the attention of walkers. Only a few centimetres in size, this amphibian spends a large part of its time in the water, sheltered from the light under stones. Its morphology, physiology and reproduction have attracted the attention of researchers, who observe it in the wild at altitude or in the CNRS underground laboratory at Moulis. After describing the revolution of the Lissamphibians and the phylogeny of the Urodeles, we will mention the main characteristics of the

Euproctus, dwelling on its embryogenesis, metamorphosis and regeneration. By carefully observing small groups of Euproctes in their tanks, we will reveal their collective instincts...

In writing this chapter, I was confronted with cutaneous respiration, which is predominant in the euproct, and this is what prompted me to write the essay on '*Low biological energies*'. A priori, the skin cells of the Euproctus would benefit from a localised respiratory autonomy to exchange dioxygen and carbon dioxide directly with its aquatic environment. I couldn't make any headway without trying to understand what was happening to the skin cells of the euproct subjected to very low temperatures during hibernation.

In chapter IV, we will look at the metamorphosis that affects amphibians, anurans and urodeles. Among the latter, the Euproctes have the capacity to pass from the larval state to the adult state, through this transition that enables amphibians, but also other animal species, to ensure a passage between two environments, aquatic and terrestrial for some, aerial for others. Using a model approach, we will go back to the most ancestral bifurcations to explore the divergent transition from the aquatic to the terrestrial environment. The processes of embryonic development, as well as metamorphosis and regeneration, are at the forefront of research in cellular genetics.

In Chapter V, let us dwell on the гёдёпёгайоп, a certain number of organisms possèdent the possiM^ of тедё^^ their organs, an opportune which is very much reduced in higher animals and in the human being in particular. It is easy to understand why research in this field is so important, in order to find applications capable of responding to the seriousness of certain pathologies, such as wounds, burns, and damage caused to cells by radiation and chemotherapy. If not to repair the aerations of nerve cells and eye damage, for example, can we hope one day to counteract the anarchic development of cells? Once again, as a naturalist, I ask more questions than I answer.

In Chapter VI, after working^ on the approach and test modèles, to try to understand the phylogènical and ontogènical aspects, then those of mëtamorphosis and cellular редё^^, our observations allowed us to reflect on the first transitions of social instincts towards a form of what is conveniently called a proto-culture. Using the concept of the reversionary effect, this opёerator proves to be very useful as a benchmark of the foundations of Humanity's Civilisation, in relation to the tendency of animals to instinctively associate in organised groupings. We have carefully observed a 'hunting party' in an Euproctes pond, to begin to perceive in a group, this beginning of transition of social instincts, which, from fagon exclusive in the human being was expressed by the civilization, of which tèmoignent the presence of the barns of Moudang, a culture of high mountain.

By way of conclusion, we will comment on biological involution and the beginnings of civilisation, describing a proto-organisation that is emerging, very distant, but pre-existing in the Euproctes, in a still haphazard association that is being determined in the incessant play of variations in a species in contact with the sèvères constraints of its environment, very changeable in altitude, from the values to the summits of the Pyrenees.

Charles Darwin's theory finds its true resonance in our exploratory approach. At the basis of his concept, it was geology that preoccupied the young scientist at his beginnings aboard the H.M.S. "*Beagle*" in the company of his fellow hydrographic officers. While carrying out surveys near the coasts, it was through the discovery of fossils, which differed in the various geological strata, that he became aware of the spatial and temporal dimension of the evolutionary process. Then, during his periods of navigation, which alternated with longer and longer explorations on land, Charles Darwin, as a very experienced naturalist, realised that it was the changes in natural environments (landscapes, climates, resources) that were so different from one another, It is the changes in natural environments (landscapes, climates, resources), so different from those on Quail Island, Rat Island, the Falkland Islands (Atlantic) and finally the Galapagos Islands (Pacific), that affect the condition of plants and the variability of animals. Evolutionary patterns that he also saw affecting the way of life of the men and women who responded to the vicissitudes of nature in Patagonia and Tierra del Fuego.

These were the beginnings of his idea, which he discussed with Captain FitzRoy during the ascent of the Santa-Cruz and on board the H.M.S. *'Beagle'*. It was by understanding the ability of animals to adapt to their environment, but above all of Thuman^ to oppose individually and collectively, through cooperation, the most acute sëlective constraints, which HE SAW during his voyage and thought about at length at Down-House.

Charles Darwin was late in adding his theory of Involution to that of civilisation when he wrote "*The Origin of Species by Natural Selection*" (1859), which led to the publication of "*The Parentage of Man*" (1972), the year in which the H.M.S. "*Beagle*" was destroyed...

Using the mediation of complex evolutionary processes applied to the Darwinian theory of "*Biological Evolution and the Transformations of Civilisation*", meticulously justified by Patrick Tort, we can probe the past to better understand the present, in order to simulate the probable trajectories of species. Will we have the wisdom to linger over the life that surrounds us and to integrate ourselves intelligently into the ephemeral vital coherences of species, in order to grasp imperatively the stochastic nature of their adaptability, where alea and chance are combined, modules that are now seriously disrupted by our socio-economic activities?

These are the patterns and rhythms that determine the future of Civilisation, the revolution of Humanity, and the survival of species and their environments, which we explored during our stopovers in the Pyrenees, to meet the Euproctes in the Moudang valley.

Chapter I

Elements of geology and biogeography, history, traditions and customs, and the civilisation of the Pyrenees.

During his voyage on the H.M.S. *"Beagle"* (1831-1836), the naturalist Charles Darwin was an expert geologist who was much appreciated by the hydrographic officers of this small three-masted barque assigned to survey the coasts of South America. The young naturalist was interested in the formation of the "fundamental" granite and the geology of the islands he explored, right up to the peaks of the Andes... During his stay on the islands that revealed the beginnings of his theory, **"from Ile aux Cailles to the Galapagos Islands"**, as an accomplished geologist, he took rock samples that were very useful for reconstructing the coastline, for which it was important to know the nature of the seabed in order to include it on nautical charts. The Royal Navy wanted to anchor its ships in a safe place so that they could drop their anchors to the bottom, without drifting towards the coast, while avoiding snagging them dangerously on a rock, or choosing to land in a port without tearing off the keel and avoiding shipwreck as they approached land. This was Charles Darwin's initial mission with the hydrographers of the H.M.S. *'Beagle'*. By sounding the coasts of South America, he established himself as an experienced naturalist, and these soundings gave him the opportunity to trace the origins of various species whose fossils he discovered with his shipmates, who were amazed by his finds buried in the strata of different geological eras. By comparing the plants and animals of the very different places in the regions he explored from the Atlantic to the Pacific, he noticed the variations caused by the different geological profiles and reliefs, which together with the climatic changes at each latitude and longitude affected the behaviour of the animals he discovered and the habits of the people he met. The naturalist observed changes in the natural environments of Patagonia, just as he did in the isolated English colonies and in contact with the 'savages', noting the living conditions of the Spanish colonists and the Amerindians in constant revolution, which he compared to those of the miserable inhabitants of the booming cities of Australia. In 1842, in *"His pencil sketch of his theory of species"*, he emphasised that geology and climate, through their changes, influence the variation of species, and he devoted a chapter to the :

"Evidence of the geology and geographical distribution of plant and animal species".

Bearing in mind his exploratory approach, for any study of the evolution of a species, such as the Euproctus in this case, I had to describe the formation and geology of the Pyrenees. We will then look at the flora and fauna of these mountains, before setting the scene geographically, historically and in terms of customs in this traditional centre of Pyrenean civilisation, described in detail by Claude Dendaletche in his naturalist's notebooks, which guided our first investigations in the Moudang valley. We give a brief overview of the formation of the I Itiutes-Pvrenees massif in order to highlight the main geological features that influenced the evolution of endemic species and influenced the socio-economic activities of this region. These anthropogenic constraints were added to the natural selective pressures of the rigours of the climate at altitude, establishing a reciprocal interaction between nature and the mountain dwellers whose study merited a few weeks in this massif to explore the upper Aure valley towards the Moudang mountains.

3.   The Pyrenees, up the Aure valley.

The formation of the Pyrenees mountain range began some 600 million years ago in the Precambrian period, when successive jolts during the orogeny of the massif led to the formation of the peaks on the Franco-Spanish border from sedimentary and metamorphic rocks, which later became the basis of plant and animal life. Later, human activity shifted to pastoralism and agriculture, followed by crafts and then industry and tourism. This land has been reworked from prehistoric times to the present day, significantly modifying the natural selective constraints at different altitudes.

The oldest deposits date from the Ordovician period (500 million years ago), and the Devonian limestones contain fossils of corals and organisms that bear witness to the beginnings of the maritime formation of the Pyrenees: a shallow sea with a coral massif on which plants and sediments were superimposed, forming the coals of the Upper Carboniferous. The Hercynian folding, which took place on these coral limestone formations until the Lower Carboniferous, changed the relief and landscape. As part of the metamorphic process, we find the fundamental granite that preoccupied Charles Darwin. This rock underwent compression and high temperatures until it gave rise to shale and marble, which are mined in the Aure valley.

The initial folding of the Pyrenean chain occurred around three hundred million years ago, during the Permian, the relief was higher than it is today, and underwent severe erosion, washing away materials and disrupting the landscape by digging trenches that filled with rock. Volcanism in turn brought its share of lava flows and ash deposits.

Red gres characterise this sëdimentation due to the oxidation of the ferruginous minerals under the effect of a then hot and humid climate.

"Iron and its derived ferric and ferrous oxides will be one of the biochemical potentials favourable to the origin of life and the revolution of species".

We have noted the presence of iron in the watercourses where the Euproctes live, and we have devoted Appendix 5 to justifying their survival in this iron exces, discussed in several chapters. In the Jurassic, dolomites[5] were formed on the reworked base of limestone muds in contact with magnesium-laden water, a potential mineral for thermal cures.

The Lower Cretaceous was a geological period of sedimentation, during which the Hercynian chain became much smaller. In the Upper Cretaceous (-100 million years ago),

---

5 This carbonate sedimentary rock is composed of dolomite containing metals that can be produced by bacteria in virtually oxygen-free conditions at low temperatures.

the Iberian and Aquitaine platforms were still separated by a maritime space; after moving apart, the plates moved closer together throughout the Eocene. In the Paleocene and Oligocene, uplift gave rise to the Pyrenees, and erosion invaded and shaped the coastline, deltas and rivers. In the Miocene (20 million years ago), rivers carried away masses of pebbles, sandstone and limestone, forming layers. In the Quaternary, glaciers and water erosion shaped valleys such as the Moudang, accentuating the isolation of the amphibians, which were subjected to strong natural selective constraints during glacial and interglacial periods.

At present, the Pyrenees cover an area of around 55,000 square kilometres, two-thirds of which is Spanish territory, and are between ninety and one hundred and forty kilometres wide and four hundred and thirty-five kilometres long. France stretches along the high, narrow northern face of the Pyrenees massif. The geological map of this mountainous territory shows us a band of primary granitic rocks, from the Mediterranean to the Pic d'Anie, which extends into the Basque country. These two homologous systems are surrounded to the north and south by sedimentary rocks, and the Pyrenees are characterised by numerous folds, a cluster of fragmented folds of hard or soft, archean and sedimentary rocks. Ahead of the Pic du Marbore (3,248 metres) and the Vignemale (3,298 metres), in the extension of the Maladeta anticline, the Neouvielle granite massif (3,092 metres) stretches between the Aure valley and the Gave de Pau. The face of the Pyrenees is a series of slopes, sliced by perpendicular valleys cut into their edges and separated by harbours below the summits. For centuries, the only way to reach these border ports was by mule track, marked by smugglers, migratory flows and invading armies.

While the climate in Spain is dry, it is more moderate and wetter towards France. In the Hautes-Pyrenees, rainfall is high from April to June and from October to November, and the Neouvielle area is very wet. Near the border peaks, climatic fluctuations are frequent, with fog and cold mists, wind and rain following one another, while the snow cover is significant and persists up to three thousand metres, giving rise to the presence of snow in summer.

Seasonal rainfall, melting snow and the Pyrenean glaciers have a strong influence on the flow of rivers at altitude. The clear waters of the Nestes and Gaves rivers rise from the cliff overhangs, cascading down rocky outcrops. The tumultuous rivers descend through a succession of deep gorges and winding valleys, flowing towards the plain where they join together in the Garonne and Adour rivers.

This geological frontier, traced over the centuries, was determined by migrations, conflicts and invasions, and by the movement of animals, livestock, villagers and their teams loaded down with raw materials and foodstuffs over the passes. In the hollows of these mountains, the valleys, places of sedentariness, exchange and passage, will acquire their linguistic identities by acquiring peasant traditions and a particular high-mountain culture, customs that will change with the industrial exploitation of wood, coal, iron and copper mines. They entered the modern era with the development of white coal, spas and winter tourism, with skiing and hiking in summer. Today, the secular traditions whose traces can still be felt are facing an inevitable confrontation with the uncertain challenges of globalisation, punctuated by advances in new technologies.

Since prehistoric times, people have decorated the caves in these mountains, and these first traces of civilisation marked the beginnings of a history that has continued from valley to valley, undergoing territorial and maritime influences at the whim of those who ruled these mountains over the centuries.

The prehistory of the Pyrenees tells the story of human and animal populations, and more discreetly that of the Salamanders and Euproctes in particular. Most of the main ornate cave sites are in the Ariege, at Niaux, Bedeilhac, Labouiche, Mas d'Azil and Gargas, whose names are evocative of the presence of humans, and to some extent Euproctes, at the foot of the Pyrenees. Humanity has left its mark on the walls of the caves with animal prints, rock art signed with its hands on the rock in the Niaux cave. As for the Euproct, swept away by the tumultuous flow of high altitude streams, it found refuge in the underground rivers at low altitude. Its artistic representation is limited to a salamander carved on reindeer antler, found in a cave at Laugerie Basse in the Dordogne....

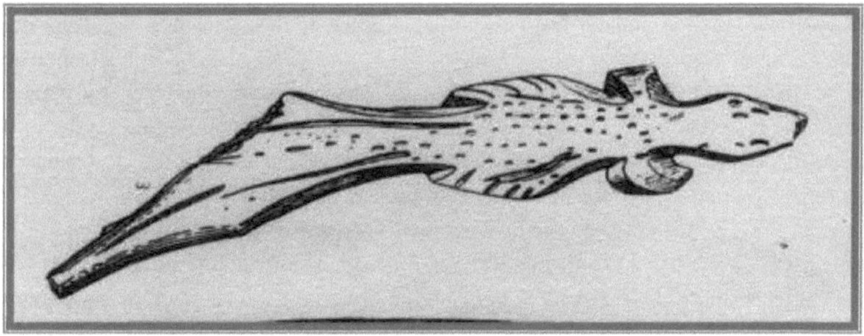

4.   *Salamander carved on reindeer antler, Laugerie Basse.*
Les Eyzies-de-Tayac, Dordogne.

We visited a part of the Niaux cave that is open to the public. Four kilometres long, this prehistoric monument consists of the famous black room decorated with black line drawings and engravings combined with furniture. In another gallery, which is not open to the public, human footprints and parietal paintings give us an idea of the fauna that its occupants hunted: herds of horses and bison, and, exceptionally, a mustelid. Paleontologists tell us that humans occupied these valleys after the Upper Paleolithic, during the Magdalenian period, between 17,000 and 12,000 years BC, after the Wurm III-IV ice age limited their presence at higher altitudes. The black room at the end of the entrance gallery must have been used frequently, and there are traces of this human presence: hearths contain burnt bones, and the floor is strewn with flint tools, including a fine bone needle used to sew beastskin garments. The drawings and engravings show that the bison is the most represented species, followed by the horse, ibex and deer. Clay engravings complete the bestiary, which includes an auroch, a rhinoceros and a salmonid, but no salamander. Near the deep gallery, children left their footprints... Hundreds of footprints and the remains of torches in the Rene Clastres network show us the path taken by the artists who drew and engraved the animals of the Middle Magdalenian period around 8,000 years ago.

The underground river at Labouiche, near Foix, can be crossed by boat over a distance of two and a half kilometres. In 1961, Euproctes were discovered there and they also live in two other caves, in the Ariege in the Siech cave (600 metres), and in another underground refuge in the Atlantic Pyrenees at the Col del Boruch. The altitude of Labouiche is only 440 metres, and the snail is not found in the surrounding mountain streams. It was carried down by the torrential waters, certainly trapped and isolated in this cave. It has become cavernicolous and shows various modifications, its depigmented skin has a dull appearance without the yellow coloration of the dorsal line, on this subject, we will return to the role of the coloured pigments of the cutaneous cells of the adult Euproctus. Its size is smaller than those at altitude and, among other physiological changes, its reproduction seems to be affected by this subterranean lifestyle, although these characteristics are reversible.

The Mas d'Azil cave is a tunnel dug by the Arize river, with a huge porch giving access to numerous galleries decorated with Paleolithic paintings and engravings, and is also populated by a large number of cave-dwelling fauna, although the Euproctus is not mentioned on this site. On the right bank, there are Neolithic, Protohistoric and historic remains (sepulchres); below the first layer, a substratum contained an abundant, well-identified fauna: ***Ursus*** *spelcus*, ***Felis*** *speleos*, ***Hyena*** *speloa*, ***Elephas*** *primigenius*, ***Rinoceros*** *Tichornus*. Traces of industry are attributed to the Aurignacian and are very significant for the Magdalenian found on the left bank. Evidence of a culture that used deer antlers to make harpoons and assegais, even going so far as to mark its presence by painting this antler on pebbles.

In the Pyrenees, parietal art gave pride of place to fauna, to which must be added anthropomorphic figures and ritualised abstract signs in a variety of forms (barbs, lines, isolated dots, lines of dots, groups of dots and lines, cupules) that show cultural progress. These abstractions were removed from the natural environment, and are the traces of an already advanced civilisation, established at the foot of the Pyrenean massif.

Among the animals usually found in Pyrenean caves are bison and horses, a few ibex, deer, a bovid and occasionally an Isard, which would indicate occasional hunts in the mountains or episodic settlement at higher altitudes, linked to a more temperate post-glacial climate.

The history of the Pyrenees does not end with the romantic story of the famous Roncesvalles Pass in Roland's Song, but is written in each valley, fixed in the urban architecture of the market towns and the rural dwellings of the mountain villages. The landscapes trace the pages of history and secular traditions inscribed in each of these valleys, a mountain culture, increasingly modelled by the uniform imprint of modernity without 6^Pёre that has developed in a few siёcles. In these enclaves lived mountain dwellers such as the Campons, Onosubates and Crebennes until the Roman occupation, which appropriated the resources of the valleys, using and abusing the mineral waters and resources of the land to then let the Visigoths and Franks occupy this territory. The Count of Bigorre often changed houses between the 11th and 13th centuries, and Jeanne de Navarre took possession of the area before it became part of the French royal domain in 1322 under Charles IV. The province was dedicated to the Count of Foix, and the Hautes-Pyrenees department was founded on 4 March 1790. Nowadays, past and present try to complement each other to keep the population active. In the 1920s, the population was around one million, and the Hautes-Pyrenees tended to be depopulated in favour of the towns. The inhabitants of the villages scattered along the valleys and waterways remained isolated for a long time, forming cantons with their own socio-economic organisations and specific customs and practices. Trails linked these natural territories, and these mosaic traditions crossed paths and met on the marketplaces of the main towns, as in Arreau, to sell produce and livestock in the presence of the bear showman.

5. A flock of sheep grazing at altitude.

In the Aure valley, farming was established on small plots of land, initially to meet the basic needs of the family, and at best, depending on the severity of the climate, surplus produce was sold on the market.

The altitude, frost, snow and hail limited the number of cultivable plants such as potatoes. Only a small variety of vegetables - turnips, peas and beans - and a few cereals (rye, oats, buckwheat and wheat) fed the families. In addition, pigs will be reared from food waste and killed on each farm, and the honey harvested will provide sugar and beeswax, enough to light the home. These crops and the rearing of cattle and sheep will require water harnessing facilities. Channelled, the water distributed by a judicious irrigation network will supply the gardens, meadows and fertile fields to obtain fodder for the long winter months preceding the transhumance to the Moudang valley. Cows, ewes, horses and donkeys marked the landscape of the Aure valley right up to the highest peaks overlooking the summer barns. Cows were used for ploughing, pulling carts and lowering the long fir trees in the mountains, which were used for ship masts, as shown in the document by Henri Gautier, an engineer from Nimes (1660-1737), on the route taken by the marine timber, which we describe in our *"Memoires de mer, cevenoles"*.

"Des cartes particulierses de la vallee d'Aure enon^ant les forêts epuisees et celles que l'on doit ouvrir, les routes tant anciennes que celles que l'on a nouvellement projettees par lesquelles on doit conduire les mats et les porter ou on les mets en radeau avec la carte du courant de la Neste et de la Garonne".

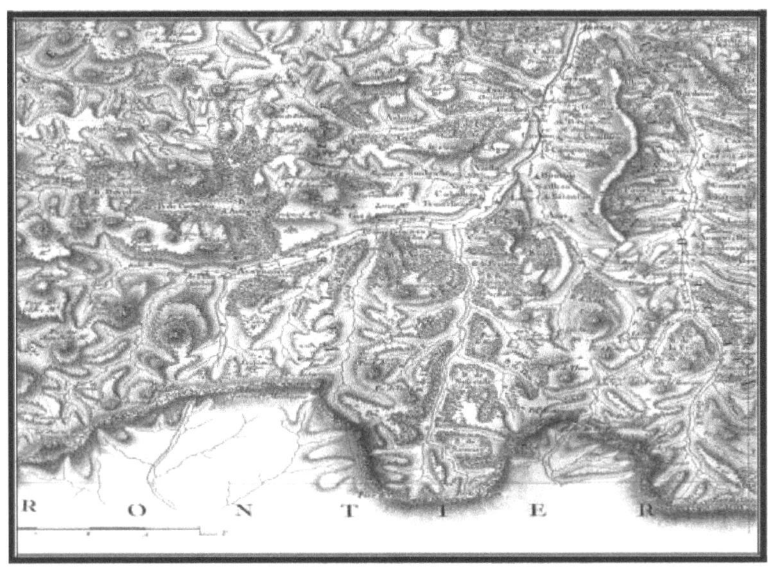

6. Extract from the Cassini map of Thury n°76/21F/Bourgoin, vallee d'Aure, from the archives of Pau, Bearn, Pyrenees.

To be shipped to the ports of Rochefort and Toulon. Timber for carpentry and joinery was an important source of income, as was timber for heating and forges.

The ice ages left geological imprints deep enough to accumulate water from precipitation, melting snow and ice to form lakes and streams that were skilfully channelled to mills. This natural water resource was then harnessed to power hydroelectric plants via penstocks. At the end of the nineteenth century, the use of candles and traditional pine resin lamps preceded the use of lamp oil, before the arrival of electric lighting. The hydroelectric industry took over, channelling the powerful waterfalls into turbines installed by mining companies in Saint-Lary-Soulan. The Lac d'Oule will be harnessed to run a dozen or so penstocks to the village of Eget, which will be totally dedicated to power generation. Most of the high altitude lakes - Cap de Long, Aumar, Auber and Oredon - will be equipped to use the water either for irrigation as far as Lannemezan or to direct it to these hydroelectric plants. This energy will enable the local population to have access to lighting and local industry to develop, and the railway lines will be electrified.

7. Source ferrugineuse de la Reine.

Once accessible, the spas in the Aure valley attracted large numbers of visitors to Arreau, Cadeac-les-bains and Tramezaigues, where the hot, sulphurous medicinal waters were used. Above the Moudang barns, in the Courroux valley (1,650 metres), the waters of the "Reine ferrugineuse" spring were recognised for their curative properties, and the medical profession became interested in its iron composition, which existed in various forms, such as carbonate, sulphate and crenate in the form of a somewhat sulphydric protoxide salt, to which a few traces of iodine should be added. Water transported in bottles lost the aggressiveness of hydrogen, leaving a ferruginous compound in the form of a protoxide salt that was sufficiently salubrious and harmless to treat anaemia and diseases that affected a depletion of the blood...

We have gathered this precious information because the Euproctes live in the larval stage, still endowed with their gills, in the highly ferruginous basins fed by these springs and waterfalls, except for the Queen's spring. This spring is the object of a mystical veneration marked by the presence of a Virgin and the frequent laying of a few bouquets of flowers in gratitude for miraculous cures. As the crow flies, Lourdes is just around the corner!

The water was good not only for curing humans, but also for cattle, which the shepherds treated at a distance of two thousand metres with a few compresses of muddy water laden with iron, the fine particles of which are deposited on the gills and bodies of the Euproctes when saturated, a problem with excess iron that I had to solve.

The protection and regulation of wildlife is shared between the National Park and the natural hunting and wildlife reserves. The Pyrenees National Park manages the Neouvielle Reserve, founded in 1936. This imposing granite massif is dotted with peaks and lakes used for irrigation and power generation. At higher altitudes, this is the domain of the hooked pine, the moors and the grasslands, which are home to a flora and fauna listed and studied by naturalists attentive to changes in these ecosystems. Winter and summer tourism involve a high level of human activity, which paradoxically needs to control its expansion while promoting socio-economic development. This means that leisure activities such as skiing, hiking and other forms of recreation must be accompanied by green leisure activities, which inevitably have an impact on the natural environment. Species such as the Desman (***Galemys** pyrenaicus*) and the Euproct (***Calotriton** asper asper*) are precious biological indicators that tell us about the quality of the water and the level of pollution in the environment of the mountains and valleys of the Pyrenees, and the Moudang valley in particular. These species help to maintain an exceptional biodiversity, which we have observed in revolutionary fashion over the last few years during our escapades, in between our exhausting cruises meant to keep the peace.

Chapter II

The Moudang valley.

After these essential gënëralitës in the Pyrenees mountains, we're off to explore the hiking trails in the Pyrenees Occidentales above the Aure valley, close to the Neouvielle reserve, which is very popular during the summer tourist season. We opted for the Pont de Moudang hunting reserve, which is better for birdwatching and easily accessible on our daily walks from the campsite we were using as a base camp. Here we are, south of Lannemezan on the northern slopes of the Pyrenees, along the Aure valley, which is still low towards Arreau, but rises from Cadeac to Saint-Lary-Soulan. Leaving this town, the steep, winding road takes us from Tramezaigue to Eget, at a place called Pont de Moudang (1,100 metres). Above the welcoming municipal campsite, the stony path that leads up to the Moudang valley can be explored on foot over a distance of ten kilometres. This itinerary leads up the Neste du Moudang on a wide track that stops on the grassy plateau of the Moudang barns, after which we have two options for progressing around the Pic de la Hount:

- Or, turn right from the barns, following the narrow path that runs alongside the Neste du Moudang, leading under the Pic de Bataillance (2,604 metres) to the port of Hechempy on the Spanish border.

- Or to the left, heading in the opposite direction, towards the Reine ferruginous spring opposite the old iron mines[6]You will then reach the port of Moudang (2,495 metres), on your way to Spain.

The Moudang hunting reserve is a very dynamic high-altitude ecosystem, but is also relatively isolated. This site poses real problems for the conservation of natural environments, flora and wildlife, which are superimposed on a large seasonal human presence, from spring to autumn. Pastoral, hunting and tourist activities take place in this area of high mountains and deep, attractive valleys. The number of hikers travelling between France and Spain is not negligible, and pastoral and hunting activities make it possible, on a small scale, to integrate the impact of socio-economic activities in order to study the evolution of species. Naturalist observations are a means of describing and measuring the impact on the biotope of a regional civilisation that has undergone major transformations in just a few centuries. The advantage of this site gave us the opportunity to move easily from the montane to the alpine and nival elevations of the peaks surrounding the Moudang barns, in order to observe variations in the selective constraints of these various high-altitude landscapes that interact on the main vëgëtal settlements and animal populations. In our notebooks, we have recorded the species identified, as well as the natural selective constraints and those resulting from human activities, compared with those experienced more intensively in the more frequented areas of the nearby ski resorts of Piau-Engaly and Saint-Lary-Soulan, as well as around the very frequented lakes in the Neouvielle nature reserve.

---

6 Iron ore from the Pyrenees was used in the blast furnaces of the Tamaris and Besseges forges in the Gard department, a use we mention in our "Memoires de mer, cevenoles".

8.   On the way to the Moudang valley.

At that time, we were beginning to become aware of the need to protect natural environments and safeguard species threatened with extinction. At the same time, pollution problems were emerging as a result of major technological disasters of international concern, as successive oil slicks hit the Brittany coast (Amoco Cadiz, 1978), and as accidents occurred in the chemical (Bophal, 1984) and nuclear (Chernobyl, 1986) industries, with worldwide repercussions, beyond national borders.

Ecology was gaining ground with the public, whose awareness had been heightened by these disasters and the highly influential broadcast of the films on *"The Silent World"* made by Captain Jacques-Yves Cousteau and his team of divers on board *"La Calypso"*, whom we frequently met at anchor in Djibouti and in the bay of Diego-Suarez in Madagascar. Jacques-Yves Cousteau (1910-1997) joined the Ecole Navale in 1930, and served on the cruiser Primauguet, the cruiser *Suffren* and the mine-hunting cruiser *Pluton*. After the cruiser *'Dupleix'*, at the Liberation, he was assigned to the underwater research centre, and became one of the inventors of the Cousteau-Gagnan regulator and the designer of numerous diving and underwater filming equipment. As a Lieutenant-Commander seconded from the French Navy, he and his wife, the true masters of the crew and divers, devoted the Cousteau team to cinematographic underwater exploration aboard the *"Calypso"* and *the "Alcyone"*. Captain Cousteau established himself as a defender of the ecology of the oceans, a filmmaker of the deep and a writer of the sea, and was admitted to the Academie Frangaise (1988). His ship, *'La Calypso'*, was a former Royal Navy minesweeper built in 1942, and he transformed it into an oceanographic exploration vessel. We saw her again swimming in a dock in Marseille (1989), then in Concarneau in 2015, restored to her original state for repairs in a shipyard. We hope that he will be able to return to sea after his problematic renovation, to preserve the memory of this maritime epic and to encourage young generations to pursue scientific research on the seas and continents.

From the events of 1968 onwards, in a context of a return to nature promoted by social movements and an awareness of the problems of pollution aggravated by major technological accidents, the nascent environmentalist ideology became clearly politicised by the action of the agronomist Ren? Dumont. The adventurer Nicolas Hulot, who was in demand as a politician, would go on to become France's Minister for the Environment. Minister for the Environment (2017), little considered, he will move on. This militant and fair-weather facade should not obscure the in-depth research carried out by scientists in a variety of specialities, whose theses and publications are little known to the public. They have long been warning of the effects of climate change and the irreversible damage it is having on biodiversity.

In 1994, Francois Ramade published an encyclopaedic dictionary of ecology that opened up environmental studies to students, extending the teaching of Life and Earth Sciences, which timidly included the Sciences of Evolution, a theory still in crisis of recognition in terms of its true biological and civilisational dimensions... Yet since 1989, in *"Les societes et leurs natures"*, the anthropologist Georges Guille-Escuret has been questioning the divide between ecology and ethnology, showing the difficulties of studying the interactions between societies and their environment. Since our meeting at the colloquium on Darwinism and society, and a few too brief exchanges before he passed away, I have kept his remarkable book on *'Le décalage humain'* (Kime, 1994), which describes the irreversible emergence of the social fact in the evolution of the human species, close at hand and available to read.

If our approach was influenced by this change in mentality towards the environment and by the day-to-day exercise of the risk prevention profession, it became that of an evolutionist naturalist, enlightened by the reading of Ernst Mayr, Konrad Lorentz, Stephen Jay Gould, Jean Chaline and Claude Devillers, but it was the final meeting with Patrick Tort that made me aware of the extent of Charles Darwin's work, which I had never known. On the other hand, my growing responsibilities in the field of safety and the environment in the National Defence and Civil Protection services led me to confront the realities on the ground in support of a theory of evolution that was undergoing as many scientific advances as it was incomprehensions about the real extent of its anthropological extension through its notorious contribution to Civilisation. It was still too much marked by sociobiology and a latent social Darwinism, as well as persistent creationist ideologies that had to be overcome in order to glimpse an intellectual openness compatible with the theory of biological evolution extended to that of civilisation by Charles Darwin, as reconsidered by Patrick Tort. At this stage, we felt it was necessary to combine the investigative possibilities offered by the considerable development of new scientific knowledge and the technical resources that could be used in fundamental and applied research. In the years that followed, I was highly motivated to develop a method for studying complex evolutionary processes, which was developed during long-term research into a probable case of ***Castor** fiber* spawning, isolated in the C?vennes. My aims as a naturalist had begun more modestly with the observation of the Euproctes des Pyr?n?es.

In 1983, as evidenced by the brief notes in the appendix, the initial aim was to carry out simple inventories as we went along, referring to the notebooks and guidebooks of the naturalist Claude Dendaletche, who drew up an impressive picture of the biogeodiversity of the Pyrenees.

In his works, he describes the geology and landscapes of these mountains, the fauna that is subject to variations in altitude and climate, and the flora that he has meticulously catalogued from the valleys to the highest peaks, dwelling in his writings on the ecology and ethnology of this Pyrenean massif. Our general observations were limited to the species we saw during the day, the Griffon Vultures, the Isards and the Marmots, and then, while keeping our attention on these populations, we paid more attention to the amphibians during our breaks at the water's edge, observing the frogs and salamanders we found in the small streams and pools that feed the course of the Neste. On our daily route, we walked up the dirt track that runs alongside the Neste du Moudang through a forest whose vegetation cover is modified by the altitude and the seasons. This little road[7] is framed majestically by the peaks whose slopes we observed attentively, after crossing a forest of fir trees and beech trees, the landscape opened onto the summits and the majestic valley of the Moudang barns. At the start of our morning route, our attention was drawn to the Pic d'Augas (2,213 metres), opposite which dominates the Pic des Aiguilles and the Pic de Born, these two peaks framing the Pic de Tramezaigues (2,572 metres) in the distance. After a few winding bends, we come across the Cuneille and Garlitz peaks (2,798 metres) before arriving on the plateau of the valley of the Moudang barns with pastures stretching across the slopes of the Soum de la Piette, Sarroues (2,835 metres) and Escalet peaks. A natural pyramid, the Pic de la Hount (2,400 metres), divides the route into two very narrow paths, while one follows the Neste under the Pic de Pene Abeilleire (2,611 metres), the other winds along the Pic des Laouas (2,381 metres) to the ferruginous source of the Reine to reach the Port de Moudang (2,495 metres).

At the Moudang barns, the bed of the Neste spreads out widely over the banks, which have been overturned by the tumultuous flow of crystal clear blue water. During the summer months, the flow is lower, and the river bed alternates between broken rocks and gravel accumulated on the earthy banks, which have been hollowed out by the current, and whose strata reveal the level reached after heavy rainfall and the melting of the snow. Around the river, with its still tumultuous flow, the meadows stretch out, covered with short vegetation and flowering plants, and the grasslands slope gently upwards. The mountains steepen towards the wooded slopes, which taper off towards the rocky summits of their peaks, with their slopes dotted with knolls and screes. Close to the barns, pastoral activities have transformed this landscape, dedicated to sheep farming, which has replaced cow farming in this valley. The flocks spread out from the fertile lawns of the barns to the higher alpine pastures, spreading out over the mountain pastures that surround these few dwellings. Set on this green plateau, a dozen or so rectangular barns, built of stone walls, are covered with slate roofs that open onto a chimney. The dark masses of the buildings, their roofs gleaming in the sunlight, cut through the vegetation. Sheltering shepherds, hunting lodge, family residence or possibly a gite for walkers, they ensure the presence of families here for a few months, from spring to autumn, before the snow falls, and are the memory of a pastoral civilisation...

---

7 Vehicle access to the Moudang hunting reserve is reserved exclusively for residents and farmers who own the barns, and for hunters during the hunting season.

9.   Barns in the Moudang valley.

The road to Moudang[8] was passable by horses and mules, and the drinkers rented the cattle barns at low prices to transport them on small ox carts from the Moudang valley to the valley of the springs at the foot of the mountains covered in birch, fir and beech trees. During the long winter season, the summer hamlet of Moudang at 1,650 metres became silent and lonely under the snow and the icy wind. In spring, the barns came alive with the arrival of shepherds and flocks, and drinkers of Chourrioux water from the iron-laden springs who came to treat their chlorosis and anaemia. They sat on the grass, ate, laughed and danced. The ferrosulphurous mineral water of Moudang was known to the peasants as a universal remedy, and the report by the chemist Filhol, director of the Toulouse medical school, described its medicinal properties (1865).

"The ferruginous water of Moudang is limpid, with a sulphurous odour and an astringent flavour, its temperature is 4°C, in the air it decomposes slowly giving a reddish deposit. One litre of water contains acids, chlorine, potash, soda and lime, magnesium (0.0030), iron sesquioxide (0.0200) and organic matter. Iron in the form of protoxide (0.0425) retains its solubility.

Stable water, he deduced, could be assimilated by the body, with the tonifying iron and the depurative properties of sulphur. Moudang water was said to cure chlorosis, passive bleeding and rheumatic pains...".

The springs[9] gushed clear and cold at a tempĕrature of 4°C and formed reddened streams that flowed into the Neste, with seasonal tempĕratures varying between - 14°C and + 34°C.

The first spring came out of a small mound of earth, fresh and pleasant to drink, with a ferruginous flavour. The second spring was more abundant, with a more pronounced sulphur taste, and no fish were to be found.

---

8 Revue de Comminges, 1888.
9 Bulletin de la societe geographique de Toulouse,1889.

10. The Neste and the Moudang barns.

These barns, known as bordes, were used to store fodder, and these small, high-altitude farms had a table to shelter the cattle and a modest dwelling for the shepherd. The thick walls were built from stones taken from the numerous scree heaps on the slopes and banks of the Neste. During construction, the walls were held together by a mortar of earth and pebbles to which a mixture of straw and cow dung was traditionally added. The walls could be plastered with lime and sand to protect them from inclement weather, snow and rain, and from the penetrating dampness of the cold fogs at altitude. The roof, built on solid sandstone beams on which the plank-covered rafters rest, was covered with lauze, and this floor was used as a hayloft. The passage of time and the various occupants have altered the exterior appearance and slightly improved the interior fittings and comfort. Some of the stonework is still visible, while on others the renovated façade rendering is discreet and compatible with the original architecture. Although the windows are narrow to keep out the cold, the entrance door was wide to allow the cattle into the barn, which is no longer the case with sheep farming, which has replaced cow farming in this valley. However, the enclosure with its dry stone walls still marks the boundary of the small property. The floor was paved with slate slate, and the simple furnishings consisted of a small shed, a solid table and wooden benches. The beds were made from a few planks of fir. The furnishings are still rudimentary, but the addition of a few conveniences has made it possible to cope with the cold. While the shepherds used to use a basic fireplace in a corner of the barn, the residents now light a real fireplace.

With one or two exceptions, the barns' hayloft is атёпадё into rooms or a matëriel shed, it serves as shelter for the summer përiode and as a convivial hunting rendezvous, much to the delight of families and hunters who reside in the valley in the right season. The Moudang barns are now a summer pasture for flocks of sheep. The shepherds live in one of these barns and look after their animals, which they park nearby when they arrive for the transhumance and when they gather them together before the descent to the valley. Usually, the sheep graze on the grassy slopes of the peaks surrounding the barns under the watchful eye of the shepherds and the guard of the dogs. Above them in flight, the griffon vultures hover and wait for one of them to stray and fall into a ravine, and then there is the "bear" whose presence is so contestëe, the dread of the shepherds of the Pyrenees...

35

11. A flock of sheep around the barns.

Atmospheric conditions are highly variable at these altitudes and conditioned our observations: visibility diminishes very quickly, it is difficult to resist freezing fog or very cold rain, and thunderstorms call for the utmost caution. As the altitude increases, these storms cause marked variations in the number of animals and people who frequent these mountains for work or leisure. From the end of July to early August, the temperature range is between 15°C and 24°C in the shade during the day, with an average of around 20°C. The wind at higher altitudes would prove to be quasi-permanent, from light to very strong, sometimes excessively cold. The daily morning mists and freezing fog at altitude became difficult to bear during long observations in the shade, forcing us to take shelter in the hollows of the protective walls of barns.

In the opposite direction, on sunny days[10] could prove to be very hot, rising to over 35°C on the overheated rocks, and becoming unbearable in the early afternoon during the long observations of the Euproctes pools during the breeding season.

12. A waterfall that feeds the Euprocte
and Neste basins
below.

We set up three observation stations, two on the Neste and one on the ferruginous source of the Reine. On the Neste, opposite the Pic de la Hount, the first station was located on a small pool fed by a stream that flowed down to the Neste, below a flooded lawn. The shallow basin, just a few centimetres long, was fed by very clear water, which calmed down as it swirled around in this natural recipient, then flowed through the green lawn before emptying into the Neste. The basin was bounded by slabs of rock and small beaches of pebbles and gravel, surrounded by larger rocks, topped by a mosaic of moss and water-soaked plants. Upstream, the second basin is larger, located beneath a waterfall that feeds it continuously, the glowing water is very ferruginous, the bed of rocks and gravel is covered with red silt, the water flows into the nearby Neste, bordered by marshy meadows.

---

10 There is an average of 1,900 hours of sunshine a year, with rainfall of between 1,400 and 2,000 mm a year, and snow cover seven months out of twelve, with average summer temperatures of between 13°C and 15°C under shelter.

The third flow we observed was from the Queen's spring, a mythical place where people believe in the "miraculous water" watched over by a little virgin in permanent bloom. This spring is very ferruginous, and we have never seen any snails in the small pool at its underground outlet. On the other hand, over a distance of twenty metres or so, highly-developed aquatic plants spread out in long filaments in the swirling water flowing towards the barns. Opposite are the entrances to former mine shafts that have been disused, and the sheep like to graze on the slopes of these hillsides, which feed the springs and waterfalls.

13 and 14, Miraculous spring of the Queen and Virgin.

*13 et 14, Source
miraculeuse
de la Reine et Vierge.*

*15 and 16, development of aquatic plants with excess iron
at the source of the Queen.*

Every day, we make the ten-kilometre journey from our base camp to the Moudang barns to set up camp near one of these pools. The approach walks take place early in the morning, when the sun is just rising, and the return walks take place late in the evening, at sunset, allowing us to familiarise ourselves with the landscape and gradually discover the flora and fauna.

As far as the flora is concerned, we will confine ourselves to describing the дёпёгаих features of the vëgëtal groupings that vary according to altitude and relief, describing more precisely the transition of the landscapes near the watercourses from the rocky systems of the streams to the marshy meadows, the wet refuge of the Euproctes.

The birds are present as we walk through the beech and fir canopy, and although their songs can be heard, it is more difficult to observe them, usually alerted by the cries of a Scared Jay that flees as we approach. When the vëgëtal canopy clears, above the first rocky summits exposed to the sun, in the warm air currents, the griffon vultures begin their circular flights at high altitude, in search of a sheep or Isard carcass. Occasionally, as we skirt a ridge, we catch a glimpse of the very rare Bearded Vulture, satisfied with a few bones for food. In the still misty meadows, groups of Isards appear, moving up the slopes towards the rocky ridges where they take refuge at altitude. In the evening, we find them in the meadows, feeding as night falls.

Every time we turn a corner, we are surprised by the whistling of a marmot. On our first trips, they were nowhere to be seen, but it seems that this population has grown significantly in just a few years. Yellow-billed ducks are constantly flying over the area around the Moudang barns and landing on the low walls. Because of the impact of domestic species and human activity on the landscape of the Moudang valley we have localisë the area of influence of each on the flora. Some horses in paddocks feed on the vëgëtaux of the meadows around the barns, while cows in other values graze at various altitudes, including on steeply sloping grassy areas.

17. Cows grazing at altitude

Sheep are distributed at altitudes of between 1,600 mëtres and 2,700 mëtres, with large flocks spread out over the heights in groups of ten to twenty sheep, starting at the level of the Neste. They appreciate the earth and rock shelters along the banks to protect themselves from the sun or rain. The contribution made by domestic animals to the ëcosystëme of the Moudang valley is major for

Involution of the vegetation cover, which is almost flattened by the cyclical passage of the herds during the day. The addition of manure by these species, combined with the reworking of the soil by their passage, contributes to the fertilisation and surface erosion that characterises the well-cleared estives up to the wooded levels of the hooked pines, and then the desertic rocky system alternating between the grassy and stony scree areas that dominate the nival level. The movements of the domestic animals are conditioned mainly by their food, which involves a progression modulated by climatic conditions (rain, sun, wind) and the need for water. The regulation of the herd by shepherds and dogs is not permanent, and there are periods of regrouping and care in the pens located on the grassy plateau close to the barns.

18. The sheep cross the Neste du Moudang.

It is clear that this seasonal presence interacts with wild species, from plants to insects, reptiles to amphibians, birds to mammals. These activities contribute to disrupting the trophic chain of the ecosystem modelled by the human presence, centred on this pastoralism and seasonal hunting, amplified by significant tourist numbers on this route between France and Spain via the Moudang valley. Walkers use the tracks and paths that they mark with their footsteps to reach the ports and summits, and the Moudang barns are a place to stop off and meet up. As a general rule, walkers do not venture off the paths, and the awareness-raising work carried out by National Park staff and the National Office for Hunting and Wildlife, as well as the status of hunting reserve, limit damage and mischief.

The teaching of Life and Earth Sciences was beginning to raise awareness of the interdependence of domestic and wild species, as well as raising public awareness of the effects of activities that affect their evolution, and this ecosystem offered me the opportunity to begin studying complex evolutionary processes by drawing up an inventory of the fauna.

19. Marmot on the lookout.

We will deal with the amphibians in the next chapter, mentioning here the most visible species that we actually encountered and we ë evoke later, the unobserved species that still persist in this environment of high mountains, we begin our description with a reptile.

Zinniker's Vipera aspis is a reptile specific to the Pyrenees. We frequently came across it on tracks and paths, on the rocks of the scree slopes near the Neste, of variable colouring, it can be identified by a dark sinuous line dotted on each side with a row of rounded spots, it sometimes measures more than eighty centimetres, its poison is very violent. The viper tends to flee when approached with caution, and can be observed at a reasonable distance, which varies according to the atmospheric conditions and the nature and location of the terrain over which it moves more or less quickly. One morning, as we were climbing the track leading to the barns, a viper came out of the ravine overlooking the Neste and crossed in front of us. Surprised, we stopped, and it continued its progress without aggression, before slipping away and disappearing into a crevice at the side of the path. During an observation of Euproctes in the basin of a stream located near a sunny rocky scree, while exploring the stream upstream, on the opposite bank about twenty metres away, a large viper was exposing itself to the sun; as we approached, it dropped heavily to the ground and disappeared among the pebbles... We periodically came across one during our explorations up to more than 2,500 metres, but we had to remain wary and avoid disturbing it when it had settled in low walls or in the overheated rocky screes that it favours, along streams, well exposed to the sun on south-facing slopes, in search of voles and other shrews. Birds dominate the Moudang valley, particularly the Chocards and Craves, which, like crows and ravens, are omnipresent at higher altitudes. The adult Red-billed Chough (*Pyrrhocoraxprothorax)* is about thirty-eight centimetres long and has a shiny black-blue appearance, distinguished by its legs and red beak, which is long and curved, more curved than that of the Yellow-billed Chough. The beak of immature Choughs differs from that of adults in its yellow-orange colour. Choughs are fond of the high mountains and do not hesitate to approach barns, walk on rock faces and find their food in the fields, search the moors for earthworms and, at higher altitudes, gather fruit and berries. Not far from their territory, on a ledge in a rocky cavity, the pair builds a crude nest of twigs and thick stems, into which the female deposits a few eggs.

The Chocard (**Pyrrhocorax** *graculus) is* almost identical in size to the Crave, with a shorter tail, a yellow beak and red legs. These two species often perform spectacular aerial manoeuvres in groups or pairs, plummeting into the air with their wings folded and attacking birds of prey that come too close to their nests. We will be paying particular attention to the birds of prey, the Griffon Vulture, the Bearded Vulture and the White-tailed Vulture, which we have actually seen in flight on the peaks surrounding the Moudang barns. During our trips to India and Africa, the Griffon Vulture (**Gyp** *fulvus)* seemed to be part of the landscape; in Africa, it is easy to observe them feasting on a carcass in the middle of the bush or playing the role of rubbish collector on a rubbish dump near the big towns. In India, the situation is a little different, as the animals are protected by the Hindu and Buddhist religions. In Bombay, at the top of the Tower of the Dead, vultures carve up the corpses in a ritual that replaces the cremation usually practised on the banks of rivers. In France, the Griffon Vulture has benefited from reintroduction in the Jonte valley, which seems to be bearing fruit with an increase in numbers. It is currently extending its territory towards Mont Lozere, where it has been frequently

41

observed by the wildlife photographer Jean Faisse. Its dispersal is being monitored, although it does not attack domestic animals directly. For the more sceptical, its excessive presence could provoke collective behaviour likely to frighten the sheep, causing defiance among farmers already frightened by the presence of wolves and bears in the Pyrenees. The evolutionary aspect of applied ecology makes it possible to deal with this type of problem. Because of the complexity of the interactions between man and animal, it is necessary to understand the consequences of reintroducing wild species over the long term.

20. Griffon vulture in flight.

In the Pyrenees, the griffon vulture, **Gyps** *fulvus, is* found in around a thousand pairs throughout the massif. It is a large bird of prey with a wingspan of two metres eighty centimetres and an average weight of seven kilograms.
The hooked beak is powerful, enough to pierce the soft parts of the skin and tear off the flesh of a sheep or isard carcass. A scavenger, it does not hunt but applies collective strategies of aërien repërage at high altitude and consumption of prey thanks to its pergant sight and flight abilities. Pairs settle on rocky aplombs and form colonies that gather in flight to fly over the crevices and take advantage of the warm updrafts around the peaks to dëplacer, without moving their wings, apart from a few movements of the digitized feathers of the wing tips and tail that regulate its planar flight. The pair incubates its young for two months in a nest clinging to a ridge, and the young vulture takes to the wing around July. At this time of year, the pair's flights with the young vulture are very spectacular. Around midday, the adults climb very high along the cliff, gliding in ascending circular trajectories in the warm currents, letting themselves fall into the void by folding their wings, then resuming their gliding flight to climb back up and repeat their spectacular fall several times. Less frequently, the pair put on an astonishing display in flight near the cliffs of the Garlitz peak: two griffon vultures come together in flight, grabbing each other by the talons and letting themselves fall in unison for around a hundred metres, before separating and resuming their flight at altitude.
From our observatory in Euproctes, we face the Pic de la Hount, whose summit is bathed in afternoon sunshine. Hundreds of griffon vultures fly overhead, hovering above the flocks of sheep, before launching into a collective curee when they find the carcass of a sheep fallen into the rocky scree.
Along the ridges, we are frequently overflown by griffon vultures and, more rarely,

approached by the bearded vulture (***Gypaetus** barbatus),* which, by its name, is half eagle and half vulture, has almost disappeared from this massif, with an estimated population of around one hundred pairs in the Pyrenees. It is the largest bird to fly over these high mountains, with a wingspan of almost two metres and an average weight of six kilograms. Its wings and tail are dark, while the ventral part appears reddish, becoming whiter towards the head, which is distinguished by eyes surrounded by dark feathers, extending into a "beard" under the beak. The beak and the rather small claws force the Gypsy to eat only bones, which it has to break by dropping them to the ground to eat them, sometimes with a little flesh on the menu. Rare to observe, it deserved a special study that went beyond the scope of our investigations limited to the Moudang valley.

The Short-toed Eagle *(**Circaetus** gallicus)* is a summer visitor to the Pyrenees, appearing almost white from below and easily identified in flight. It has the particularity of feeding on reptiles, which it hunts with its keen eyesight. It is protected from the bites of venomous snakes by the hard hairs around its beak and its very long legs, covered in a thick scale, which extend into rough talons with which it captures viperids without risk.

These birds of prey play a fundamental role in the food chain and in the natural regulation of species, without any harm or nuisance being attributed to them. As soon as a wild or domestic animal is spotted, the crows surround the crushed carcass in a ravine, then come the vultures and the Egyptian vultures (not observed) which fly over the site and land cautiously to fight over the remains, and finally the lammergeier comes to swallow a few bones.

21. Isards, Rupicapra** pyrenaica, **on the heights.

With a little care, we can spot Isards perched at altitudes of between 1,100 and 2,300 metres, most often seen in the morning and evening when they come down to feed in groups on the grassy slopes near the barns. This ruminant bovid is known as the rock goat, **Rupicapra** *pyrenaica, and is* smaller than the Alpine chamois. Its coat is more distinctive in winter, dark brown on the body and turning beige on the shoulders and thighs, with black markings on the chest and muzzle. Its hooves cling to the rock, the snow on the walls and the steepest cliffs, which it scrambles over in search of food. It has a fairly large stomach and its blood is rich in red blood cells. It can live for around ten years, although it is predated by eagles and foxes, and hunting it is a veritable institution, if not a well-established tradition in the Moudang barns. We often see it on the rocky heights overlooking the valley, in the shade of a pine tree or standing out against a snow bank.

On our first visit, we didn't come across any marmots, although a few years later they were a daily sight. Marmots have been reintroduced to the Pyrenees and their dispersal seems to be very prolific. This large rodent, almost a metre long including its tail, weighs up to four kilograms in late autumn and two kilograms in late winter, after hibernation. Dark brown to lighter in colour, its fur is greyer on the head and tail. Its elongated snout is distinguished by a black nose extended on either side of the top of the head by the eyes and small ears, and its jaws bear strong, continuously growing incisors that it wears down as it feeds on roots and plants. The most frequent sightings were on the grassy and rocky slopes lining the paths, where a shrill whistle signalled their presence and alerted their congeners. To withstand the cold of winter, the marmot hibernates in its burrow, gradually falling into a state of lethargy by lowering its breathing from thirty to one inhalation per minute, with the heart rate dropping from one hundred to forty beats per minute, which increases the carbon dioxide content and reduces the internal heat from 36°C to 5°C. Below this temperature, there is a risk of loss of vital functions, in particular the brain, which periodically reacts to this limit by re-establishing the metabolism. The animal wakes up every three weeks to carry out its excretory functions before plunging back into winter lethargy, curled up in the dry grass of the burrow. Between March and April, the marmot colony wakes up, emaciated, and slowly resumes its activities before the breeding season.

This period of hibernation, which we develop at greater length in our research and in the essay devoted to low biological energies, seems to us to be of particular interest for studying reversible biochemical processes at the cellular and organismal level, in particular the role of mitochondria and brown adipose tissue in animals that hibernate or live in cold environments. These processes become rarer, if not non-existent, in higher mammals and man in particular.

22. Pair of marmots and lookout on a rock.

Now that we have described the species that are frequently seen, we need to talk about those that we did not encounter, and in particular those that are the subject of virulent debate and controversy, starting with the brown bear (***Ursus** arctos*), which numbered almost two hundred individuals in 1937. At the time of our first visit, there were still around twenty bears in the Pyrenees, including the famous Cannelle, who was shot in 2004, after Melba had been massacred in 1994. Three bears of Slovenian origin were reintroduced to the Pyrenees, a female in 1996 and a male in 1997, and although their numbers were small, the population reached over fifty after these highly contested reintroductions. The presence of these bears is a real source of fear for angry farmers, and arouses just as much passion among ecologists. We are faced with a problem of society and biodiversity that will have to be resolved, just as with the wolf, which is naturally aggressive in its quest for food. We believe that the scientific and technological conditions are ripe for preserving species while at the same time promoting socio-economic activities. According to the shepherds we met, whether at Pont de Moudang or in other valleys, there are indeed attacks linked to the presence of the bear, which kills a few sheep and sometimes causes several dozen of them to fall into the ravines, which inevitably stirs up their anger, periodically triggering passions that manifest themselves in a certain aggressiveness and conflicts with its ecological protectors. The bear usually lives in its territory in the Neste d'Aure, at high altitude, where it finds sufficient vegetation, insects and rodents to feed on, possibly chasing wild ungulates, and more easily, trying to capture domestic animals by making murderous incursions at low altitude.

On the other side of the world, conservationists are expressing their concern for the conservation of species and biodiversity, and in the face of these concerns and hopes, wildlife managers and public authorities are trying to find solutions to the problem we encountered in the Himalayas, from Nepal to Bhutan. In these mountains populated by thousands of bears (**Melursus** *ursinus* and **Ursus** *thibetanus*) we observed the same problem, but managed differently depending on the region of the Himalayas where they are protected. The precepts of Buddhism consider animals to be sentient beings, and they are not hunted. In the highlands of Bhutan, which has introduced a policy of protection and very controlled exploitation of the forests, the inhabitants have to contend with a number of threats, not only from tens of thousands of bears, but also from felids (**Panthera** *tigris*, **Panthera** *pardus*), as well as occasional attacks on their herds by the Snow Leopard (**Uncia** *uncia*) at high altitude or the Valley Leopard (**Neofilis** *nebulosa), and* a multitude of macaque monkeys and langurs, which pest the plantations. As a result of the attacks on the herds, goat farms, whose wool was used to make scarves, have closed down. The government has had to encourage younger people to take up this type of farming by leveraging the flourishing tourism industry, which is reasonably limited by the exorbitant prices of travel to Bhutan. We observed small isolated farms in the mountains surrounded by a very dense forest occupied by numerous animals including monkeys and bears. The property, gardens and animal pastures are surrounded by a system of fences to protect them from intrusion by monkeys looking for vegetables and fruit, and the almost constant presence of humans and dogs helps to keep them away, particularly when a group of monkeys invades a field. For the bear, any unexpected encounter is dangerous, and the shepherds, who bear large scars on their faces, report this presence without animosity. They are armed with long knives worn on their belts and stay alongside their herds with dogs to counter bear attacks. At high altitude, Snow Leopard attacks on livestock are rare, although this high-altitude feline does occasionally attack young yaks in the pastures.

23. In Bhutan, on a high-altitude farm, hedges and human surveillance protect against invasive Grey Langurs, Semnopithecus *entellus.*

In Nepal, the problem is different in the Himalayan zone, from the Kathmandu valley to the Terai, where the forests are intensively exploited to make charcoal close to the towns. Wildlife varies from the northern foothills of the Himalayas to the plains of the south, on the Nepal-India border, where nature reserves help to protect wildlife and combat poaching. In the south, the tiger's survival depends on the protection of these last natural areas between India and Nepal. This particularly dangerous species requires vast territories and a food chain that is subject to the pressures of hunting and poaching, which are difficult to control. On the other hand, socio-economic priorities, poverty and latent corruption do not make protection easy. While in Bhutan, local efforts and culture make the presence of wild animals such as the bear acceptable, in Nepal, outside the reserves and national parks, their fate depends on the possibilities of rural and urban development supported by international organisations whose programmes consist of meeting the basic needs of the inhabitants while safeguarding the natural environment. The participation and interest of villagers in the conservation of species and the environment are, along with education, a way of preserving the biogeodiversity of these areas and creating jobs to help them escape from poverty.

In comparison, the problem of reintroduced bears in the Pyrenees seems to us to be easier to solve if we are willing to look at the local history that has marked this territory, and study its phylogeny, ethology and ecology, in order to place its conservation in a long-term evolutionary context, which can only be complex, meaning that any solution will have fairly unpredictable and undesirable responses. It is clear that the concerns of the herders must be addressed; we have shown that in the Moudang barns they have shaped the mountain through their vital pastoral traditions. For centuries, the Pyrenean ecosystem included the presence of bears, hunted and dominated, before their near extinction. The reintroduction of bears taken from Slovenia is still too risky to satisfy farmers and ecologists.

We are faced with a choice of society that is effectively the result of complex evolutionary processes in biological evolution and the transformation of a civilisation that we can summarise as two complementary solutions to favour biodiversity. The context of socio-economic development, in clear opposition to natural environments, rules out short-term solutions, which are generally emergency measures involving the eradication or relocation of dangerous individuals. The aim is to safeguard a dozen species of ursid in the world and many other more or less dangerous species that are really threatened with extinction.

- The first phase, in the medium term, consists of creating an integral reserve made up of islands to preserve a sufficient population to guarantee the genetic potential of a vital territory, free from all human activity. This requires an in-depth study of the reintroduction areas gradually created in mosaics, using the most effective scientific and technological resources to guarantee the autonomy of the reintroduced species in these survival islands. This means ensuring the integrity of the sites, with control centres to guarantee public safety with the help of multi-disciplinary scientific research, as well as public information, relying imperatively on the essential consultation and participation of local stakeholders. There are many experiences (Lion in Africa, Tiger in India, Panda in China) from which it is possible to draw inspiration to apply them by атёНогап^ this also concerns the urgent safeguard of the rimurians of Madagascar, the monkeys of the old and new continents and especially the great apes of Africa in danger of extinction, the last tëmoins of the phylogënie of Primates.

- The second phase is a longer-term meeting with the active human population and the public to carefully integrate the reserve into the local socio-economic dynamic, by judiciously extrapolating the high-predation intégrales reserves to low-predation rural conservation areas, extending as far as the urbanised përiphëries, judiciously rendered inaccessible to the most aggressive animals. It's a very delicate and complex management, out of an unstable equilibrium, but essential to re-establish and maintain a biodiversity compatible with social and economic activities that will find themselves valorisëes in the long term.

Firstly, there is the action of safeguarding a species threatened with extinction, which requires the impërative reconstitution of a ëcosystëme cohërent of predation, the territorial^ of animals (bear, wolf, tiger, ëlëphant) has no frontiers apart from that of the territories they set for themselves and that our civilisation would like to attribute to them. Other species in the Pyrenees, which are certainly less dangerous, should be included in these biodiversity enhancement initiatives: the Desman, the Badger, the Otter, the Voles and Bats, the Ermine, the Squirrel, the Rock Ptarmigan, the Greater I^tras and among the birds of prey, Percnoptere and Golden Eagle, Great Horned Owl, Tengmalm's Owl, not forgetting the atypical Bec-croisë des sapins to name but a few, among the multitude of birds, insects and plants that can be seen in the Moudang valley.

The discreet Pyrenean skink is not as demonstrative and aggressive as the so-called wild species that have left a lasting impression on the memory of the local inhabitants, more so than the modest and discreet batrachian of the Urodeles family that we observed along the Neste du Moudang.

Chapter III

The Pyrenean skink.

The Euprocte des Pyrenees belongs to the order URODELA (Dumez.il. 1806) of the family SALAMANDRIDAE (Goldfiss, 1820) of the genus Calotriton (Gray, 1858) and the species *Calotriton* asper (Duges[11], 1852).

24. Red frog and Pyrenean snake.

In the Pyrenees mountain range, the alpine elevation limits the presence of high altitude plant and animal species, in particular the Lissamphibians Urodeles and Anurans. Their distribution in these high mountains is affected by a harsh climate, with snow and cold reducing their active life to the short summer period at the hottest temperatures, during sunny days. The Batrachians include tailless anurans such as the Red Frog (***Rana** temporaria*), which is widespread in Europe and in the Pyrenees above 500 metres, and the Natterjack Toad (***Alyte** obstetricans), which* can be found at altitudes of up to 2,400 metres. The Urodeles with tails include the Tritons and Salamanders, which belong to the Salamandrid family.

---

11 D'apres les Urodeles de France, Annales des Sciences Naturelles, Zoologie et biologie animale, volume 17, page 253-272.

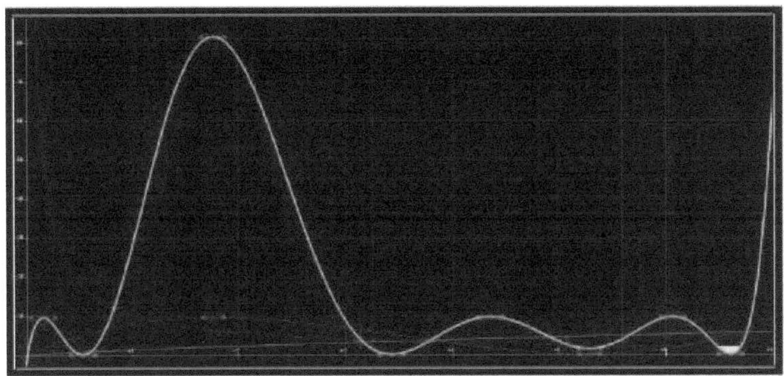

25. Diagram of the evolution of the Urodeles branch (red curve, green arrow), its wavelet transformation (green curve) gives a peak of strong expansion, followed by a regression with successive modulations, the coefficient of adaptability is estimated at 0.28 (shaded green area on the right) for present-day Urodeles, i.e. an evolution with a stabilised asymptotic trend (blue curve).

26. Pyrenean skink,
Calotriton asper asper.

These include the endëmic Euproctus of the Pyrenees. **Calotriton** *asper asper* (Dugës, 1852) is a close relative of the Tritons, the Corsican **Euproctus** *montanus* and the Sardinian **Euproctus** *platycephalus,* with which it is often compared, but whose phyla are thought to be different. There is a clearly different species in Catalan Spain, **Calotriton** *arnoldi,* found in the Montseny massif in Catalonia. **Rhithrotriton** *Derjugini Nesterov,* discovered in Kurdistan, **is** thought to resemble **Calotriton** *asper asper.*

The Euproctus is found in the Spanish and French Pyrenees, with its presence noted at altitude, in the highly oxygenated waters of the torrents and lake basins of the Hautes-

Pyrenees, as well as lower down in the rivers and caves of the Ariege. One of the first to report its presence, Louis Ramond de Carbonnieres wrote a letter to the explorer and naturalist Alexander Von Humboldt (1769-1859) on the fauna of the Pyrenean lakes (1821).

"At Lac d'Oncet at 2,314 metres, at the foot of the Pic du Midi, there are no fish left, but there are aquatic salamanders. I don't remember seeing any above 2,518 metres, at Lac du Mont Perdu...".

In 1895, the naturalist Jakov Von Bedriaga devoted lengthy studies to the Euproct and described a more massive variety with developed tubercles. We should also mention the studies by the Italian zoologist Michele Lessona (1881) on the histology of the skin of Corsican species and by Roule (1909) on the teguments of the Pyrenean Euproct. Finally, we should mention the research of Lapicque and Petetin (1910) on the apneumic respiration of a Corsican specimen, which is comparatively partial in the case of the Pyrenean specimen which Camarano (1896) considered to be intermediate between the pulmone state and complete apneumia.

In 1923, the Euprocte des Pyrenees was studied by Professor Raymond Despax, who reported its presence in a number of lakes.

"The animal is abundant throughout the region close to the Neouvielle".

Nowadays, **Calotriton** *asper asper* or Euproctus, is found in the streams near these lakes, whose waters flow down the slopes of the mountains surrounded by alpine meadows interspersed with forest cover which, towards the summits, give way to rocky screes. The flow of these high-altitude streams varies from high to very low, and some experience periods of seasonal freezing and drying. These climatic and ecological conditions, as well as human activity, are selective constraints that should be considered as important adjustment variables for Euproctes populations in the Pyrenean mountains.

Its range extends from the Basque country to Catalonia, and it can be found in the Pyrenees Atlantiques, Haute-Garonne, Ariege, Aude and Pyrenees Orientales, from 200 metres to almost 3,000 metres. Researchers are studying Euproctes in the wild, as well as in the CNRS underground laboratory at the Theoretical and Experimental Ecology station in Moulis. The biologist Franeois Gasser observed it at the biological station at Lac d'Oredon (1,849 metres), and university laboratories are studying its development, metamorphosis and organ regeneration. For scientific purposes, scientists have studied the Axolotl, **Ambystoma** *mexicanum,* in particular, in the laboratory. This is a bred Mexican form that is now protected from all experimentation. In comparison, the Slovenian cave-dwelling **Proteus** *anguinus* has characteristics more specific to subterranean life (appendix 4).

During his in-depth research into the development of the Axolotl egg, Professor Jean Signoret (1931-2007) demonstrated the Blastulean transition (1971). He worked in the nuclear physics laboratory of Louis Leprince Ringuet (1901-2000), a native of Ales in the Cevennes, before heading the developmental biology laboratory at the University of Caen. Professors Jean Signoret and Jacques Lefresne have described in detail the Blastule transition during the early segmentation phases of the l'reuf of l'Axolotl.

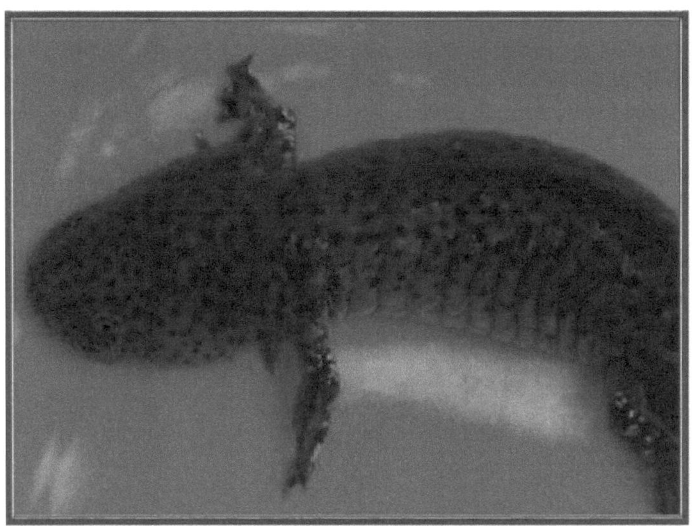

*27. Axolotl, **Ambystoma** mexicanum*

Fundamental and applied research into cell development and regeneration has opened up therapeutic avenues for repairing damaged tissues, with the long-term aim of finding solutions to serious and degenerative diseases. Our revolutionary knowledge of amphibians, Urodeles and Euproctes in particular, has encouraged us to gain a better understanding of their complex evolutionary processes by seeking to identify the causes of their adaptability to altitude and cold, taking into account their ability to lead an aquatic and terrestrial life.

The origin of the populations of Euproctes that are distributed in the watercourses of each of the valleys of the Pyrenees dates back to the Tertiary period, at the end of the Ice Ages, as a result of climatic fluctuations and geological upheavals, they would have been geographically isolated by a separation of the populations that evolved in the lakes and watercourses of the valleys of the Hautes-Pyrenees. The evolutionary study of the Urodeles and the Euproct in particular has attracted our attention because of the relative phylogenetic stability that has affected this species for sixty-five million years, while taking into account the adaptations that distinguish them according to altitude, notably for reproduction, metamorphosis, and more particularly the constitution of their skin, which is very involved in cutaneous respiration, not forgetting their exceptional capacity to regenerate their organs.

Euproctes des Pyrenees populations range in altitude from two hundred metres in the Corbieres Audoises to two thousand five hundred and sixty metres in the Hautes Pyrenees, with its maximum frequency estimated at two thousand metres. Its discreet presence is a good biological indicator of the quality of the water in torrents and rivers, as well as the state of their surrounding vegetation and mineral environment.

It is a stënothermic animal, living in cold water at a temperature of no more than 13°C. It is rheophilic, with an attraction for oxygenated running water, and stënophobic, avoiding sunlight, although the presence of euta pigments indicates that its body is not insignificantly photo-sensitive.

His experimental orientation to magnetism (Schlegel, 2006) involves all his senses and organs sensitive to light, his eyes, his pineal gland and his skin with its pigments. It has a very clear positive stereotactic effect, remaining flat against the background. There are morphological and physiological differences between populations, but these are very slight and reversible. The most visible morphological difference is in the development of skin tubercles and yellow to orange patches on the skin.

27. Calotriton asper asper.

According to F. Gasser, protein studies show affinities with current Asian forms, some of which are giant, such as the **Andrias** *japonicus* salamander in Japan, which is two metres long and weighs thirty-five kilograms.

What we have here, then, is a relict species from the Tertiary period that has taken refuge in the Pyrenees massif and, by splitting up its living space, has been sensitive to successive fluctuations in cooling during periods of glaciation and to post-glacial warming, a succession of geo-climatic and ecological fluctuations that have enabled it to selectively settle and establish itself in these mountains without any major changes over the millennia.

The Euproctus[12] The average adult euproct measures between ten and fifteen centimetres and weighs around ten grams. Its body, which is more or less grey-brown to black, has a granular appearance and is extended by a laterally flattened tail and well-developed legs. It has four toes on its front legs and five on its rear legs, and moves by contortion, propelling itself underwater. At the surface, it crawls along the bottom of the water,

---

12 Protection and status in France: The Euproctus is protected by the decree of 22 July 1993, articles R
211/1 of the rural code, which prohibits its destruction, mutilation, capture, removal, transport, sale,
naturalization and removal of its eggs.

clinging to the rock with its clawed feet.

Its very rough skin is dotted with small asperities made up of cellular aggregates with horny tips, in the centre of its back, on the axis of the spinal column, there is a yellow pigmented line[13] by lipophobes, while the underside of the body is more or less orange to vermilion in males.[14] for males. The back is grey to greyish-black in colour, with melanophores contributing to the changes in colouring. The whole body absorbs different wavelengths of light and heat. The larval stage is dark and shiny, and sexual maturity is not reached for four years, depending on the climatic conditions that limit its development. The Euproct practises hibernation, remaining motionless and without feeding; during this period, respiration is very slow and is exclusively cutaneous, and it cannot tolerate negative temperatures; after a state of lethargy due to hypothermia, the action of the cold, at temperatures below 0°C, causes irreversible lesions. During hibernation, the membrane of cells and their organelles is composed of a lipid bilayer that is fluid at normal temperature. At 0°C, this fluidity is maintained by the action of unsaturated fatty acids containing proteins, which temporarily prevent a fatal phase transition, although it freezes below 0°C. As for borderline hyperthermia, this seems to be around 37°C, a temperature common in full sunshine on the rocky slopes bordering streams and pools where Euproctes take refuge from the sun.

The male has a prominent globular cloaca, split at the base of the tail, while the female's cloaca has a conical appearance. Breeding begins in spring, from June, and continues until August at 2,000 metres. When the water warms up to between 12°C (active spermatozoon) and 18°C, at the right temperature for mating, the male and female coil around each other for long hours during the amplexus. When their cloacae are close together, the male emits between one and four spermatophores containing spermatozoa, which, if not carried away by the current, will fecundate the female's ova. Depending on the weather conditions, egg-laying takes place four weeks after fertilisation, but can be delayed by a year from the time the snow melts until late summer, if the temperature conditions are too low for the eggs to develop. When the eggs are laid, the female reaches a rock shelter in the water and deposits the second eggs, averaging five millimetres in diameter, in isolation under a stone. The young larvae hatch at around four weeks, remaining in this larval state for a year or more, depending on altitude.

We observed one of these tiny, shiny black larvae emerge from the water in a small pond, climb a small vertical rock face a few centimetres above the water and move, contorting itself on the overheated horizontal rocky surface, towards the shelter of mosses and plants that were still damp. This terrestrial phase is short-lived, before returning to the small pools to feed. In this larval state, gill and cutaneous respiration predominate, pending metamorphosis, to which we will devote a chapter.

The adult euproct's head is flattened, its snout is rounded and it reveals globular eyes with eyelids and a circular pupil. Accommodation of vision to the water and air environment is achieved by deformation of the crystalline lens and displacement of the eyeball in its orbit. As a result, when the euproct closes its eyes, they form a protrusion in the mouth, a feature that helps it swallow its food and contributes to bucco-pharyngeal respiration. It has a

---

13 Pteridine Xantophores are poorly soluble in water and fluoresce. This aromatic electron acceptor compound is thought to be involved in hydroxylation and oxidation during skin respiration in its water-soluble form as riboflavin.

14 Carotene erythrophores vary from orange to red during the breeding season.

large jaw with S-shaped teeth, and its tongue, which is fairly advanced for gripping, enables it to capture and ingest a variety of prey, such as molluscs and aquatic crustaceans. It feeds on fly larvae, mosquitoes and ephemera, and hunts the many insects that fall to the surface of the water. Capture is not very selective; floating and drifting in the current, live prey are snatched up and jerked into the jaws, which crush them lightly to hold them in place. Euprocte do not chew them, they are swallowed whole, not without difficulty, depending on their size and resistance, and oral absorption is slowed by the convulsive movements of their limbs, which shake to neutralise the insect. Thus absorbed, the prey progresses through the digestive system under the effect of gastric contractions at an optimum air temperature of between 15°C and 32°C during the day, for an acid pH of 6 in the stomach. If the temperature is too low, the puffer reduces its hunting activity and thus its diet, making it more difficult to digest prey and slowing down its metabolism. When the temperature falls below 6°C, the adult Euproctus leaves the water and takes refuge in a rock shelter that is quickly covered by snow, then hibernates on land. Its metabolism is reduced to a few heartbeats and lymphatic crevices, which enable it to survive until spring, when it prefers streams and small pools to feed and reproduce in the water.

At low altitude, this occasional cave-dweller remains trapped in underground water, isolated from other populations, and has acquired characteristics specific to life in the dark and exclusively aquatic, which influence its development into a giant larva and the slowing of the light-sensitive metamorphosis process. In these conditions, it has a diet adapted to cave-dwelling fauna.

Compared with fish, they have a larger brain and their head can be oriented using an atlas vertebra. To hunt in the water, they use their sense of smell rather than sight, and are sensitive to vibrations in the aquatic environment via receptors in the lateral line. From the outside, the nostrils visible on the snout open into the nasal cavity via the choanae located on the roof of the mouth, which has mucous membranes lined with olfactory cells.

They work just as well in water as they do in the dry, by adapting the sensitive cilia of the cells, which lengthen in the open air above the mucus layer, whereas in water the cilia are retracted below the mucus layer, which is reduced to a less active water film.

The longevity of the euproct is estimated at twenty years, and morphological and colour differences can be observed from one valley to another, even between rivers. Breathing takes place via several routes, gills and skin in the larval stage. After metamorphosis, the adult still has cutaneous respiration, supported by the bucco-pharyngeal and reduced pulmonary respiratory systems.

28. Euproct larva with gills.

In a liquid environment, the larva has three pairs of gills, which disappear during metamorphosis. In the adult state, pulmonary respiration takes the form of a saccular cavity with thin walls formed by folds that extend into pulmonary alveoli. These lungs are not very efficient and play a part in regulating the buoyancy of the euprote. Muco-pharyngeal ventilation predominates over that of the lungs, the thin mucous membranes are highly vascularised and breathing is regulated by movements of the floor of the mouth, which compensate for cutaneous respiration.

Cutaneous respiration takes place in the open air via the skin, which is moistened by the thickness of the mucus, and when immersed, directly in contact with the water. Compared with pulmonary and oral-pharyngeal respiration, cutaneous respiration is permanent and is thought to predominate over the other two.

The circulation дёпёгаle of the blood, dёroule under the action of the creur, in arteries, capillaries and veins that vascularise the organs and the skin, draiiK'e by the lymph. At the larval stage, the heart is similar to that of a fish. The adult has a globular creur, with two

55

auricles and one ventricle, the red blood cells in small quanta are nuck^s. 1. lK'moglobin of the Euprocte presents a к'ПиИё for oxygen inferëior to that of mammals, it is between 7.5 g/l and 15 g/l depending on the altitude and turbulence of the watercourses. The lymphatic system, which irrigates the skin cells in conjunction with the gënëral circulation, is therefore essential for skin respiration. The direct supply of water through the skin allows the diffusion of dioxygen for respiration and the evacuation of carbon dioxide, as well as the localised excretion of metabolic waste products. A dozen or so lymphatic chambers, consisting of reservoirs animëzed by rhythmic contractions, drain the lymph into the venous system, which contributes to osmotic balance and cutaneous respiration, when it absorbs water through the skin. The kidneys have iK'phrons similar to those of fish, they ё^^к large quantities of water and cause the retention of salts. As the Euproctus does not drink, in the event of drought or extreme cold, salts are secreted while water retention remains very limited, which would increase its resistance to cold during hibernation.

Its transition from water to land requires us in the following chapters to develop successively the modal^s of its triple respiratory system, by exploring its phylogënie and the initial conditions of its ontogeny as well as the process of mëtamorphosis, conditioned by altitude and the pvreiK'en climate. Its respiratory system gives us an initial insight into the chronology of the evolution of the Lissamphibians Anurans and Urodeles, whose variations through discrete mutations will be subject to strong natural selective constraints that appear to stabilise them.

Chapter IV

The breathing of the Euproctus, between water and land.

To computer this description of ***Calotriton** asper asper*, we need to delve deeper into the role of its respiratory systems for атёlюгег coefficients of biological adaptability and model divergence in order to devalue the probabil^ devolution of this species that lives between water and land. The Euproctus adapts practically to each valley it occupies by adjusting its physiology and metabolism to local conditions within the narrow limits of temperatures that fluctuate according to altitude and season. Professor Raymond Despax (18861950) of the natural history laboratory of the Toulouse Science Faculties and director of the hydrobiology laboratory at Lac d'Oredon was very intrigued by the reduction in its lungs and studied the respiratory behaviour of the adult animal. In the bulletin de la Sociëtë Zoologique de France, at the sëance of 26 May 1914, he concluded:

- Although the lungs were small and functional, he did not consider pulmonary respiration to be essential.

Pursuing his observations:

- He noted that cutaneous and oral-pharyngeal breathing were sufficient to ensure vital respiratory exchanges. By noting, bucco-pharyngëe respiration is of minimal importance and it is in the absence of pulmonary respiration that,

"Skin breathing plays an essential role".

This system of breathing through the skin is essential to its aquatic life and to its survival in hibernation[15] on land, and is probably an adaptive remnant of the extremely cold climatic conditions of the Ice Ages, followed by warm, no less disturbing, post-glacial climates. These cyclical climatic episodes and gëological upheavals proved to be very selective, and many species, and the Lissamphibians in particular, got round them by undergoing mëtamorphosis, an extremely complex adaptive process that gave them access to the terrestrial environment, although it was still necessary to understand how this evolved.

Professor Raymond Despax has demonstrated the very active contribution made by the cells of the tëgument to cutaneous respiration in the adult Euproctus, whereas in the case of aquatic larvae, respiration via the gills was predominant until metamorphosis, complementing this ancestral respiratory process. Long adapted to alternating hot and cold periods, this method of diffusing dioxygen directly through the skin cells favoured its transition from the aquatic environment to the sufficiently humid terrestrial environment.

---

15 During hibernation, frogs possess a tolerance to hypoxia and acidosis, at an ambient tempërature close to 0°C under the control of a very simple central nervous system, respiration becomes slow, if not ëpisodic.

29. Euproctes basin, below a waterfall.

The paper on Raymond Despax's highly methodical experiments on skin vascularisation showed that the superficial teguments of the epidermis and the deeper teguments of the dermis play the role of a veritable respiratory and excretory organ that is locally autonomous. The skin of Urodeles in general and of the Euproctus in particular is richly vascularised by three complementary pathways, that of the lymphatic vessels on the one hand, and that of the arterial and venous capillary networks of the general circulation on the other. This descriptive and experimental demonstration reveals three physiological possibilities that can be selected during the Lissamphibian revolution. These are respiratory preadaptations that are expressed to a greater or lesser extent in Anurans and Urodeles. In the course of their evolution, as an initial condition, their dominant cutaneous respiration is incidentally supplemented by oral-pharyngeal respiration, giving rise to the pulmonary system. These three modes of respiration exist simultaneously in the Euproctus, but with different degrees of efficiency, as indicated by R. Despax in his histological study of the teguments of the skin.

"The epidermis is thick, with a horny layer made up of two layers of cells, beneath which are five layers of epithelial cells covering the connective tissue of the dermis.

He began his study with a description of the tëgument, the roughness of which is due to the distribution of tubercles over the whole body, conical prominences, horned at the top and coloured brown to black. Their external cutaneous arrangement can be verified by observing the surface of the body of **Calotriton** *asper asper*, and their size varies, becoming more massive from the head to the tail. The description highlights other small protuberances as well as the fine excretory ducts of the skin glands and tubercles. The thickness of the epidermal and dermal layers of the skin varies along the body, as do the protuberances of the various tubercles and glands. R. Despax gives a description of the skin cells and how they are vascularised, the epidermis is made up of :

- The superficial layer of flattened, horny cells is relatively constant in thickness throughout the body, but thickens markedly around the tubercles.
- The intermediate layer is made up of three to four rows of cells.

Attentive to the slightest detail, R. Despax noted the spatial arrangement of the cell nuclei, and in this respect, the cells of the epidermal epithelium show a microadaptation to the predominantly aquatic environment[16] . A non-negligible sensitivity to external conditions caused by the establishment of a differential pressure between gravity and the water push exerted through the cutaneous tissue, this water gradient acts by diffusion into the cutaneous tissue and osmosis through the membranes of the stratified cells of the epidermis. This adjustment is manifested by the polarisation of the cell nuclei, an arrangement which probably has consequences for their intra- and intercellular functions, according to R. Despax. Despax, their effective exchange cross-sections in contact with the cytoplasm differ according to their position in depth:

- The cells of the deepest layer have an oval-shaped nucleus, the long vertical axis of which is directed towards the surface of the skin, while the opposite axis is directed towards the base of the connective tissue, which is bathed in lymph, the primordial role of which we shall emphasise. As the largest effective section of ovoid is oriented vertically, these elongated nuclei, which are highly polarised laterally, offer a large lateral surface area for exchange with the cell cytoplasm.

- The nuclei of the cells in the intermediate layers are on average more or less spherical, depending on their position.

- For the upper layers of epidermal cells, the nuclei located in the lower layer of the horny layer are flattened, with the largest surface facing the external water environment. They have a highly directional horizontal effective cross-section, directed respectively towards the external environment (water, air) and the interior of the cells. This suggests that the epidermal cells have an affinity for capturing dioxygëne from water and releasing CO2. The variations in thickness are greater for the intermediate and deep layers of cells than for those of the horny layer.

The dermis has a very variable ë thickness, it is endowedë with a йогтё tissue of connective fibrils impregiK'es of lymph, pigments are localised at its upper part and the cuta^s vessels form a very spëcific vasculo-pigmentary system. The separation between the epidermis and the dermis is not regular, capillaries surrounded by a thin conjunctive envelope, originating in the dermis, propagate into the epidermis. Beneath the epidermal cells, the connective tissue of the dermis, oxygenated by cells with directional nuclei, is often thicker than the epidermis, and in particular the space between the skin glands is surrounded by a vascularised connective envelope.

---

16 In comparison, we will comment on the adaptations of amphibians in space in the chapter devoted to the development of the miif.

30. Detail of epidermal protuberances.

In an aquatic environment, this provision makes it easier to
-    The localisëe oxygenation of the lymph, in conjunction with that of the arterial and venous capillaries of the gënëral circulation of the blood.
-    The evacuation of carbon dioxide (CO_2) and cellular waste (NH_3) is said to be the result of localised detoxination, which takes place directly from the connective tissue to the epidermal cells under low external pressure, and externally via the lymph, towards the sëcrëtric and excretory cells grouped together in aggregates in specific glands, simultaneously vascularised by the arteriovenous system of the general blood circulation.
In the terrestrial environment, Гохудёпайоп is mëdiatisëe by the thickness of the mucus and the excretion dëvoluted to the glands would be partially dërivëe to the bloodstream дё^ток and the renal systëme (uric acid) by a functional adaptation to altitude, related to preadaptations to the aquatic environment. This dërivation ëtablishing a progressive adaptativ^ of the respiratory and excretory systems to terrestrial life, during involution of the Urodëles.
In terms of evolution, cutaneous respiration is still essential. For this purpose, the skin has a vascular system with a high content of highly active respiratory and non-respiratory pigments, positioned at the boundary between the dermis and the epidermis, according to R. Despax.
"The combined presence of pigments and blood vessels is very abundant.
Sometimes crossing this boundary, the capillaries push in front of them a thin layer of connective tissue associated with a non-iK'gligettable spread of implanted nerves into the thinning epidermal epi^lium, the blood capillary no longer passing through the skin surface, but exclusively through the stratum coriK'e with its three layers of flattened cells.

These thin exchange surfaces help to promote the absorption of dioxygëne dissolved in water and the direct release of carbon dioxide, as well as other more or less toxic substances produced by the cells and stored in the mucus and toxic glands.

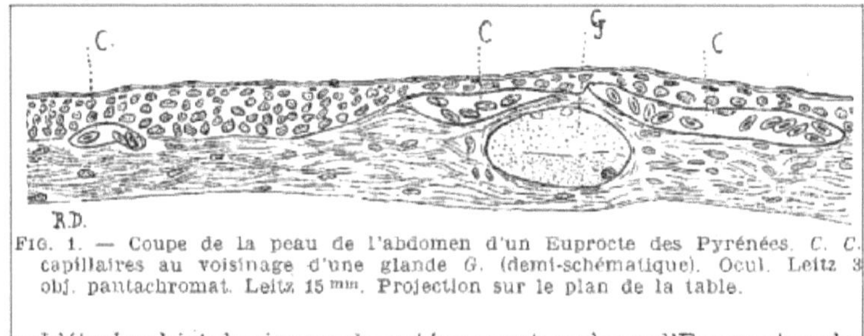

FIG. 1. — Coupe de la peau de l'abdomen d'un Euprocte des Pyrénées. C. C. capillaires au voisinage d'une glande G. (demi-schématique). Ocul. Leitz 3 obj. pantachromat. Leitz 15 ^{mm}. Projection sur le plan de la table.

31. Diagram by Raymond Despax, epidermis and dermis of the skin of the Euproctus.
On contact with the air, as a result of a reaction of the skin cells, the nervous system and the hormonal system, which are sensitive to an increase in temperature and variations in light, a protective mucus is formed on the tëgument, originating from certain protruding sëcrëtric and excretory glands.

In water, the cells of the epidermis, connective tissue, lymph in particular, and blood capillaries are directly supplied with dissolved dioxygëne. Respiratory pigments of moral origin come from ferruginous waters, these oxygen collectors sensitive to solar radiation and ultra-violet rays at altitude, are located superficially between the epidermis and the dermis. Under these optimal conditions for diffusing and capturing dioxygëne, according to R. Despax, cutaneous respiration takes place mainly on the large exchange surfaces of the body, on the abdomen, head and tail.

We'll be coming back to this cutaneous respiratory system, and in particular the major role played by lymph in conjunction with the mesh-like distribution of the arteriovenous capillaries of the general blood circulation around the excretory canal of the toxic mucus-secreting skin glands.

Following this detailed description, R. Despax carried out experiments in the presence of Euproctes, progressively depleting the water of its oxygen.

- In normally oxygenated water, he observed a movement of the floor of the mouth which caused water to enter the buccal-pharyngeal cavity. Breathing would then be partly buccal-pharyngeal by absorption of the dioxygen dissolved in the water, diffused through the connective tissue of the buccal space, which extends into the saccules, of which he suspected a very limited pulmonary use.

- By observing it underwater, we have noticed that it periodically releases small bubbles. We will see that the management of the volume of water sucked in and expelled, containing air and carbon dioxide, has consequences for its behaviour in the water, particularly when it voluntarily expels water and gases from its primitive lungs to regulate its buoyancy in order to capture insects. Should this device for adjusting buoyancy be seen as a preadapted and selectable respiratory system that is functionally more efficient in anurans?

- As it emerges at the surface, the euproct swallows a mouthful of water and continues its rhythmic breathing by absorbing atmospheric air, which has a higher dioxygen content than water, so that cutaneous respiration loses its effectiveness. To do this, it makes an extra effort by using the bucco-pharyngeal system, extended by the primitive lungs, which incidentally fill up with air. Could the response to this respiratory stress be facilitated by the effects of solar radiation at the surface, which would have contributed to the internal contraction of the ear by deforming the floor of the mouth, triggering bucco-pharyngeal respiration to compensate for the oxygen deficit in the open air?

- Respiratory exchanges with the skin are indistinguishable in both cases, which in principle does not prevent this process from being functional, provided that more mucus is produced in the open air to avoid dehydration and continue to favour respiratory exchanges with the skin when in contact with humid air during terrestrial progression. If the temperature rises, under the effect of stress, the nervous system reacts and causes mucus to be emitted, reaching a maximum thickness. At the limit of adsorption by the mucus and the skin, the need for dioxygen places greater demands on the bucco-pharyngeal mode, which is extended by the pre-adapted lungs, which become selectively more functional during the course of the revolution.

With reference to the phylogeny of the Rhipidistians (Appendix 2) and the revolution of the Urodeles outlined in the diagram in the previous chapter, with respect to the formation of the lungs[17] how did such a chronology of adaptability of these different respiratory devices become established in the Euprote?

If we go back to the initial conditions of the Rhipidistians and Tëtrapods to the Lissamphibian Urodeles, we can see in the examples below a progression of respiratory systems during the transition from an aquatic to a terrestrial environment.

From the Upper Carboniferous to the Lower Permian (-280 Ma to -270 Ma), **Seymoura Baylorensis** (Broili, 1904) was a reptiliomorph amphibian some sixty centimetres long that had the appearance of a salamander, its skin considered 'dry' indicating a terrestrial life in an arid environment.

The Houiller et Permien d'Autun-Epinac basin[18] described by Henri Emile Sauvage (1890) revealed the presence of **Protriton** *petrolei* (Gaudry, 1875), a larval specimen with gills and "soft" skin, whereas **Pleuronoma had** "thicker" skin. The Permian basin at Lodeve contains numerous marine and freshwater lacustrine fossils that are extremely useful for analysing the transition from amphibians and reptiles to the terrestrial environment.

At the time of the Lissamphibian revolution, in terms of assimilating the parameters of their adaptability into models, the fossil evidence for cutaneous respiration is very limited, and cutaneous tissues are not preserved, apart from any armour, dermal plates or protective scales. This means that we need to take into account variations in other morphogenetic factors that are cautiously correlated with changes in cutaneous respiration, in particular the progressive adaptations involved in skeletal strengthening, reproduction and the development of metamorphosis. These abilities are essential when ecosystems change, bringing with them natural selective constraints, to which we now have to add anthropogenic pressures.

According to our observations during the summer, Euproctes intensively hunt insects that float on the water. To capture them, they make numerous movements under and on the surface, then dive, dragging their prey to devour it in one of its rocky shelters. In addition,

---

17 Salamanders from the Plethodontidae family and the Telmatobius culeus frog with its many folds of skin (Lake Titicaca, 3610 metres) have no gills or lungs.
18 Geological synthesis of the French Permian basins. BRGM,1989.

the reproductive act is long and exhausting, and during the intense efforts of hunting and reproduction, skin respiration may not be sufficient to obtain enough dioxygen. Initially, this gene would motivate the adults to seek out the highly oxygenated tumultuous waters of mountain streams. Over the longer term, in the course of their evolution, in response to this respiratory gene induced by such sustained efforts or in reaction to related aggressions, predation, interspecific competition, climatic, geological and ecological fluctuations, the vital oxygen deficit became more pronounced, first underwater and when swimming at the surface, then incidentally in the open air where oxygen levels increased. Faced with these situations of intense aggression for the organism lacking vital dioxygen, the Euproct's nervous system would have engaged various pre-established respiratory compensation processes. The processes of dioxygen assimilation and skin excretion would have been progressively adapted from the lymph to the general circulation of the blood (creur, kidneys), following a selective sensitivity to the change of environment, in particular bucco-pharyngeal respiration, which became more efficient in the open air.

Considering R. Despax's experimentation in support of our observations of the hunting and reproduction periods, we suggest that the ancestors of the Pyrenean Euproctes have amplified the instinctive bucco-pharyngeal contractions that they use to swallow their prey. When they swallow a large insect, it is the alternating sinking of the eye sockets into the buccal space that leads to the movement of the buccal-pliaryngeal floor. These contractions, which cause the insect to move forward, engage the entry and exit of the water, and allow them to ingest a good proportion of air incidentally when swimming close to the surface. This situation of respiratory compensation and behavioural overcompensation has proved selectively favourable to the development of pre-adapted respiration, thanks to the presence of connective tissue conducive to the absorption of water, the diffusion of dioxygen and the release of carbon dioxide through the buccopharyngeal cell mat. As it gradually took hold, this respiratory rhythm, under the control of the nervous and hormonal systems, supplanted the more discreet rhythm of the locally pulsating, autonomous skin cells that take place in the lymphatic creurs, complementing the general circulation of the blood, which is regulated by the creur's beat. More solicited in the open air than in the water, bucco-pharyngeal respiration, by progressively forging the air towards the two poorly developed pulmonary saccules, tends to be selectively as favourable as cutaneous respiration, without however predominating the latter in the Euproctus.

In continuing his experiment, R. Despax had artificially depleted the water of its oxygen. This reduction in oxygen levels had occurred naturally during the Tertiary period, as a function of altitude, during alternating periods of excessive cold and heat. The Euproctus, probably subjected to an accumulation of stressful situations, instinctively favoured taking in air at the surface via the bucco-pharyngeal system, which had become sufficiently efficient thanks to this respiratory adaptation that stimulated the saccules of the developing lungs. There would have been a connection between the localised cellular pulsations of autonomous cutaneous respiration and the cadence being acquired by mouth-pharyngeal respiration associated with the use of the future lungs. The contractions produced in the lymphatic creurs are linked by a differential coupling with the general circulation, which is regulated by the frequency of the creur's beats, as confirmed by his observations.

"When the floor of the mouth is lowered, air enters by depression into the nostrils,

after a few oral-pharyngeal respiratory pulsations, the nostrils close and the glottis opens, the floor of the mouth, as it rises, pushes air towards the lungs, oxygenated blood from the general circulation is propelled by a heart with two auricles and one ventricle".

If we were to favour this evolutionary hypothesis, the multiple stresses that would have involved this bucco-pharyngeal compensation, accentuated by overcompensation at the limits of vital coherence by acting at the level of the primitive lungs in the making, would have resulted in an instinctive behaviour of flight to survive a persistent respiratory gene. Under highly selective environmental conditions, with frost and dryness, would its ancestors have tried to leave the water to become terrestrial for a while, to hibernate or to try to reach another salutary watering place, favourable to the populating of the rivers in each valley?

Taking into account the fortuitous biological variations caused by the rare mutations and random genetic derivatives of a small population subject to the effects of the random constraints of natural selection, there would have been complementarity between the three respiratory systems, with the beginnings of the establishment of pulmonary respiration still reduced to its simplest expression, usable until then for buoyancy. This primitive mode of respiration more or less supplanted bucco-pharyngeal respiration adapted to the hydrous environment, which had incidentally become compatible with open-air respiration, still supported by the archaic cutaneous respiration that is still very effective to this day in water, and in air under certain fairly strict conditions of temperature and humidity... In his physiology course (1848) at the Faculty of Medicine in Paris, Pierre Berard wrote.

"Breathing takes place on contact with the skin, and this exchange of gases constitutes a genuine respiratory process involving the absorption of oxygen, the release of carbonic acid and the exhalation of nitrogen. This cutaneous respiration supplements that of the lungs, particularly in cold-blooded animals such as amphibians.

Their naked skin is used for aquatic respiration, and temperature influences the life span of lung-deprived amphibians. Spallanzani noted that submerged frogs lived longer in winter than in summer, at around one degree they lived for eight days and died in a day when the temperature rose above two degrees".

In his experiments on toads and salamanders, he found that, at a temperature of $0°C$, their lifespan tended to triple, whereas at $32°C$ they survived for just 12 minutes. Yes, it is true that cold water contains a greater quantity of dissolved air and dioxygen, accentuated by the mixing of the water. As I explained in the essay on low biological energies, we also need to refer to the quantum effects of which I describe the 'Cold' solutions and then the thermo-hydro-dynamic 'Hot' solutions. At low temperatures, these effects act on the autonomy of the respiratory capacities of skin cells in an aquatic environment, and on land, particularly during hibernation, within narrow limits of viability depending on the species.

"These quantum cold solutions, for $T\text{-}o < 0°C$, are characterised by a high entropy and a small variation in energy A e as strictly nuclear free energy, which tends towards a maximum of low-entropy energy E at $-0°C$, to rise to the thermo-hydro-dynamic regime at $T > -0°\,C$ to $+0°C$".

In his course in medical zoology (1869), Louis Roule described the cutaneous respiration of several animal species, the respiratory process of which consists of liquid osmosis in an aquatic environment, or gaseous osmosis in a terrestrial environment, as the absorbed dioxygen and the carbonic acid released by the cell pass through the tissues. Gaseous osmosis occurs through thin tissues, with the animal breathing through its skin over the entire surface of its body, a direct exchange process that initially seems to be organised more specifically in the digestive system of certain species. In the Tunicates and Vertebrates, the respiratory organs are located in the anterior region of the digestive tract, whereas in the Echinoderms[19]In the Echinoderms, from which today's Echinids derive, which have no respiratory apparatus, a specific portion of the intestine, through its wall, allows respiratory exchanges with the water that circulates in the intestinal cavity towards the apical pole. In Enteropneustes, the anterior region of the digestive tract is pierced by a series of openings, and the water that the animal absorbs by mouth circulates through canals and exits through pores. The portions of tissue separating the canals are criss-crossed by numerous blood vessels, and the blood absorbs the dioxygen from the water by osmosis through the walls of these canals. In tunicates, a fold of the tëgument surrounds almost the entire digestive tract, thus transforming it into an internal gill, a specific location for cutaneous respiration. Gills develop in the larvae of Anurans and Urodeles, providing respiration in conjunction with the skin before metamorphosis. In terms of adaptability, Urodeles have benefited from random morphogenetic variations and natural selective chance events, with structuring and de-structuring effects, through their metamorphosis, which favoured a physiological phase transition between the aquatic environment and the open-air terrestrial environment, favouring cutaneous respiration, gills in the larval state and bucco-pharyngeal respiration in the adult state.

In Lissamphibians, this adaptive context will allow a change of environment, as increasingly developed respiratory preadaptations progress at the speed of natural selection, causing a convergence towards pulmonary respiration, which will become sufficiently developed for many species to adapt to terrestrial life. During its reproductive period, the male alpine newt (***Triturus*** *alpestris*) develops a small dorsal ridge that increases its surface area for cutaneous respiration, and it effectively uses mouth/pharyngeal respiration by contracting the floor of its mouth. Like the latter, the Pyrenean snail, which is still very aquatic, has maintained a four-breathing system, including the use of gills in the larval stage, before metamorphosis. However, for this species in the adult state, cutaneous respiration still plays an essential role, which raises a number of questions. In particular, the function of epidermal cells, which vary in thickness and shape in contact with the aquatic or terrestrial environment, depending on altitude and climatic conditions (Appendix 2). Without neglecting the decisive role of the connective tissue of the dermis, intimately associated with the nerves, lymph, blood capillaries and photosensitive respiratory pigments of the epidermis, this cutaneous system seemed to us to be an essential problem that needed to be studied in greater depth for several reasons. In particular, for those relating to its development during embryogenesis and during the metamorphosis of ***Calotriton*** *asper asper,* whose organs regenerated after an incidental amputation - a process of unexpected regeneration that has raised questions among many researchers!

---

19 Echinoderm fossils with limestone skeletons are discussed in the volume entitled "Les fougeres noires", which is devoted to plant fossils from the Carboniferous Cevennes.

In the adult euprotect, the skin is naked, viscous and moist. As we have just described, it is made up of several layers of cells with polarised nuclei which are formed during embryonic development and metamorphosis. Skin respiration takes place during gaseous exchanges via epidermal cells in direct contact with water, and by necessity during terrestrial life, in the air; this exchange takes place via mucus, the thickness of which is conditioned by humidity and external temperature, depending on altitude, seasons and climatic fluctuations. The dermis is the deep layer of connective cells with an elastic texture. The skin has the particularity of not accumulating reserve materials and of evacuating the waste products of its cellular metabolism largely locally. It adheres only at a few points, which serve as passageways for the nerves and blood vessels that supply it with blood and lymph. The epidermis is the visible tegument, its cells ensuring exchanges with the outside world in contact with water or air, in particular the secretory glandular cells incorporated into the dermis, whose prominent excretory orifices face the outside of the epidermis. Uniformly distributed throughout the body, in the air, these functional cellular aggregates secrete the fluid, transparent mucus that forms a hydrophobic film of variable thickness.

In a terrestrial environment in contact with the air, this film facilitates skin respiration by limiting evaporation as a function of variations in temperature and humidity, which fluctuate with altitude and harsh seasonal atmospheric conditions. Under the skin, the same mucus glands or more granular specialised glands produce and excrete a toxic liquid. This venom repels potential predators, but above all, its initial function was to :

"Local detoxification of the skin and skin cells involved in skin respiration.

Water is absorbed into the skin by simple osmotic diffusion, with the dioxygen dissociating from the water in the cells of the epidermis and the connective tissue of the dermis, and finally being absorbed successively by the fine lymphatic vessels towards the network of arterial and venous blood capillaries. Between the dermis and the epidermis, various respiratory pigments participate locally in the capture and transport of oxygen, which is then redistributed throughout the body via the general blood circulation. If we consider this localized, practically autonomous cutaneous respiratory system, the lymphatic circuit would behave like a network in the opposite direction, determining the oxygenation of skin cells compared to the general blood circulation with its low level of red blood cells. These haemoglobin-bearing red blood cells have an iron atom to capture the dioxygen from the more or less ferruginous water captured by cutaneous and oral-pharyngeal respiration. These oscillators, particularly iron, are active locally as electron transporters, whereas pulmonary respiration[20] is almost non-existent in the Euproctus.

Let's open a first parenthesis on this dioxygen trap. Ferrous iron (*Fe++*) was probably present in the primitive oceans (*Haldane, 1955*), and is found in suspension and at the bottom of the highly oxygenated ferruginous pools of the Euproctes of the Pyrenees. Ferrous iron *(Fe++ ions)* is oxidised to ferric iron *(Fe+++ ions)* by oxygen, giving rise to an insoluble rust-coloured ferro-ferric hydroxide, an iron oxide magnetite with strong permanent magnetisation (*de Duve, 2011*). The quantum field effects of this biochemical oscillator on the atoms and molecules in its immediate vicinity cannot be ignored. This reversible biological oscillator, present in hemoproteins and chromoproteins containing

---

20 In anurans, the vascularisation of the skin is more specifically provided by the blood circulation, which is generated from the pulmonary arch to a cutaneous artery.

metallic elements, gives off electrons. The transfer of these negatively-charged particles (ions, electrons) and positively-charged particles (protons) contributes to skin cell respiration in the mitochondria. Does the magnetic field of these metal ions affect the spins, ultrafine structures and chirality of the atoms in the skin cell medium, nucleation and the formation of aggregates?

To answer this question, I devoted an essay to the low biological energies involved in the cell's metabolism. A deep dive into the abyss of quantum physics was needed to explore the abiotic role of this excess iron and of the ferro-oxidising bacteria in the watercourses and ferruginous basins where the Euproctes graze.

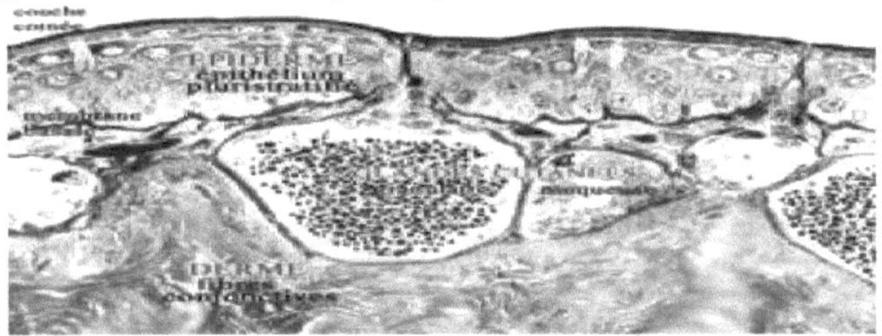

**32. *Cross-section of the skin of a Urodele (Echalier G, 2002), the skin is bare,
the epidermis is the superficial epithelium,
the connective tissue of the dermis is irrigated by lymph.***

The thin cornëe layer covering the epidermis is a very perrëable cellular base, and is the site of the osmotic exchanges that contribute to cutaneous respiration, "rëgënërëe", and is eliminated during moulting. Urodeles have mucous and granular glands, the number of which varies according to altitude and environment.

The mucous glands located beneath the epidermis sëcrëtent a fluid, transparent mucus; the skin is protected by this hydrophobic film of variable thickness, which contributes to cutaneous respiration and helps maintain the internal osmotic balance during aquatic life, especially during the days on land.

The granular glands, which vary from species to species, have been spëcialisedë in response to multiple stresses. In the event of attack by a predator, they produce a thick liquid rich in venomous substances, and are thought to have the ability to protect the skin from fungal infections and bacterial infections.

Urodeles are more or less coloured depending on the presence of carotenoi'des[21] depending on the season and age, particularly during the more pronounced breeding period. The dorsal and ventral teguments carry chromatophores, which are proteins of various colours that may contain a metallic element. These pigments are distributed in three layers between the epidermis and the dermis, and are sensitive to solar radiation and ultraviolet rays. Melanophores from brown to black produce their own pigment, melanin, a complex polymer derived from tyrosine metabolites, and their stimulation by light tends to darken the skin. This darkening is widespread on the dorsal part of the euprotect, increasing the amount of light and heat absorbed, protecting against UV rays and contributing to thermal balance. There are yellow xanthophores, such as the line running

---

21 Carotenoids are hydrocarbons that are not very soluble in water and are associated with the chlorophyll in the lipid layers of the chloroplasts that contribute to plant photosynthesis.

down the back, which made me wonder about their photosensitivity, and erythrophores, which are orange to red pteridines that cover the belly. Guanine guanophores, the purine base of DNA and RNA, degrade to produce urea and allantoic acids, which are released into the secretory glands and the general bloodstream.

*33. Adult euproct, dorsal line
of yellow xantophores*

These pigments can migrate to the surface as mëlantophores, which darken or lighten the tëgument depending on luminosity and tempërature. A drop in temperature causes darkening, while desiccation and a decrease in luminosity act on lightening as in the case of occasional or permanent cavernicoles. This set of pigments is sensitive to external influences and to internal changes during the reproductive period, which is controlled by the nervous and hormonal systems, and, depending on age, reacts to various stressful situations. Their role as mineral germs, as activators (guanine) of cell development, particularly during the regeneration process, which begins with the activation of interstitial stem cells, remains to be demonstrated.

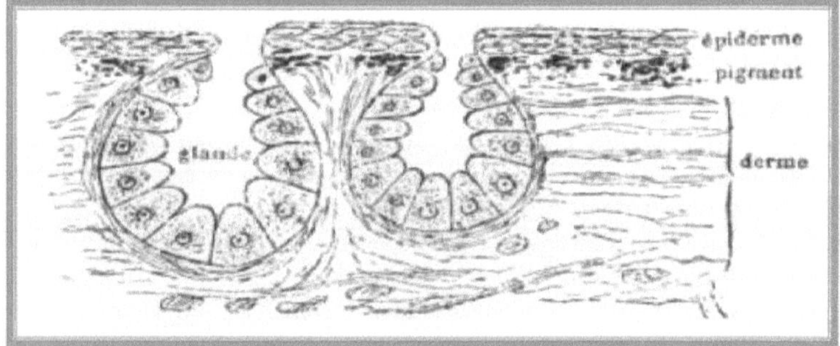

*34. Summary diagram of the skin: In relation to the external
and internal environment, the
skin of the Euproctus consists of
an epidermis stratified into two or three layers of cells in relation to
the connective, muscular and nervous dermis,*

68

There are indeed pigmentary clusters between the epidermis and the dermis, chromatophores which we suspect play an active role in cutaneous respiration and cellular development by nucleation and aggregation. In this highly dynamic area of the skin, there are also interstitial cells that remain in the state of stem cells, which can be activated under certain conditions. These cells will be called upon during metamorphosis, and more specifically in the case of tissue regeneration, they may be recruited to reconstitute an organ that has been accidentally amputated.

This pustular-looking skin, which is smoother and more viscous in the larva than in the adult, plays two essential roles: - *As the main respiratory apparatus in contact with the water*, supplemented by buccopharyngeal respiration stimulated by the passage from the aquatic environment to the open air, which is itself extended by the incidental use of two alveolar saccules used as residual lungs.

- *A local secretory and excretory apparatus*, whose detoxifying function we believe to be very important in justifying the skin's cellular autonomy.

What is the role of this excretion function, as well as that of localized dioxygen uptake and the evacuation of carbon dioxide by the cells, in organ regeneration?

This question invites us to continue exploring the role of the skin and its components, successively in water and in air. In Lissamphibians, the lymphatic vessels are highly developed, and together with the blood vessels of the general circulation, they branch out and form networks that open into large canals. The thoracic duct, dividing into two branches at the front, sends chyle and lymph into the anterior venous trunks. At certain points, lymphatic reservoirs are animated by rhythmic contractions and constitute what are known as lymphatic creurs. In frogs in particular, two of these are located in the scapular region under the skin of the back and two others behind the hip bones.

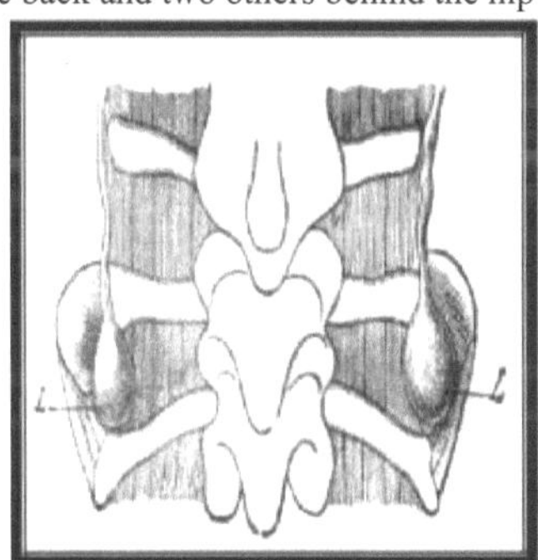

35. Lymphatic ca'iirs (L) of Rana esculenta.

The lymph and the network of arterio-venous capillaries that irrigate the connective tissue and sëcrëriice glands are involved in cutaneous respiration and local detoxification by this double differential drainage in the opposite direction. The entire amphibian lymphatic système appears as a branching tree terminated by networks or terminal ampullae, bounded by an endothelial wall.

Its blind ends, anastomosing into a terminal network, plunge into the connective tissue of the skin and mucous membranes close to the blood vessels of the venous system and the nerves.

69

In ***Rana*** *temporaria*, posterior lymphatic crevices are formed by a thickening of the wall of a venule, among Urodèles, the formation of these lymphatic crevices was studied by Zacwilichowski and Grodzinski (1917) in the Newt, ***Molge*** *vulgaris*. They described the first appearance of these creurs as a thickening of the venous endothelium. In 1926, Grodzinski, on the Axolotl larva, observed the development of valvular creurs on the venous wall, followed by the appearance of myofibrils that produced the contractions of the lymphatic creurs by expelling the lymph towards the venous system, forming pulsatile reservoirs, the lymph would be admitted into them when they dilated and it would leave when they contracted. In the archives of microscopic anatomy and experimental morphology, Justin Jolly describes his research into the lymphatic system of batrachians. He considered the lymphatic system to be a set of firm vessels with a very dense network extending into the connective tissues of the skin and mucous membranes, without merging with the blood vessels. Biologist Justin Jolly attributes the following roles to the lymphatic system:

- It would supply the blood with cells elaborated in the lymph nodes, to this end, I develop further the nucleation and aggregation process on the cellular modèle of the water drop.
- It transports nutrients.
- It is a reservoir of water for the blood and contributes to internal balance.

Its lymphatic crevices are thought, on the one hand, to have a venous origin by budding, and on the other hand, to have independent vascular germs participating in the elaboration of the lymphatic system. In these cases, I wondered about nucleation processes initiated by vortexes on ion-germs (pigments) up to the formation of aggregates capable of engendering such a generation or provoking regeneration from an active interstitial stem cell. In Anurans, lymphatic creurs collect lymph from subcutaneous sacs, which Justin Jolly has described in detail; the frog has four lymphatic creurs, while Urodèles are thought to have a greater number; there are up to fifteen lymphatic creurs in the Salamander larva. He mentions the rhythmic contractions of these creurs, some of which regress during metamorphosis, probably during the predominant establishment of the general arterio-venous blood circulation.

In the essay on low biological energies, we defined the fundamentally random causes of pulsations in the inter-membrane space of the mitochondrion by means of a proton field pulsed to saturation, which at threshold would coordinate cellular activity, producing the energy (ATP) needed to induce the contractions of the myofibrils that give rhythm to the pulsations of the lymphatic creurs and the main creur.

For an Anoure, at a tempĕrature of 20°C, the pulsation ratio of the creurs would be 75 pulsations per minute for the lymphatic creurs against 29 pulsations per minute for the creur of the gĕncralo blood circulation, i.e. a multiplication factor of 2.58 which corresponds to a linear regime with damped fluctuations.

If the contractions of a dozen or so lymphatic creurs in the euphroct are more numerous than the pulsations of the creur of the general circulation, is the autonomy of cutaneous respiration confirmed at low temperatures, increasing survival and longevity?

Cutaneous respiration is particularly effective during hibernation, proportionally, when the heartbeat decreases to one pulse compared with five contractions per minute, in favour of the lymphatic creurs. The pulsation of the main heartbeat is of high amplitude, whereas the successive contractions of the underlying lymphatic vessels are of low amplitude and are prolonged between two systoles of the heartbeat of the general circulation, whose long diastole is discreetly supplemented by the five contractions per minute of the lymphatic vessels. The creur beats would depend on the threshold pulse field resulting from the mitochondrial network of the cells whose ATP would activate and coordinate the movement of the cardiac myofibrils. If we consider the hypothesis that the activity of the creator cells is coordinated by this resulting field, is there a discrete 'Cold' solution in the quantum domain that would produce a minimum free energy in a 'Hot' thermo-hydro-dynamic regime, sufficient to maintain life at around 0°C?

The mitochondria absorb or store sodium in the *Na+/Ca* exchange^{++} which regulates the concentration of extra-cellular calcium in the myofibrils, conditioning the contractility of the main creur, supplemented by those of the autonomous lymphatic creurs. The imminent role of c_{4++} will be seen in the chapter on egg development, where contact between the spermatozoon and the oocyte causes a calcium discharge that depends on mitochondrial activity.

Depending on the atmospheric pressure at altitude, the height of the rivers and the buoyancy adjustments linked to the behaviour of the skimmer, the water that diffuses through the skin cells acts on the circulation of the lymph, which carries nutrients and cellular products, hormones and the white blood cells of the specific immune systems. The action of the lymph and its creators is therefore not to be overlooked in the nucleation and formation of aggregates, during a scarring process and the regeneration of an amputated organ, insofar as there has been a prior integration of these processes in the genes of the skin cells during the evolution of the Urodeles.

As the skin has no reserves, the lymph drains excess fluid and helps to deoxidise the skin via the epithelial cells, in particular towards the secretory and excretory glands. The mucous glands that secrete mucus and the scaly glands are embedded in the dermis; the scaly glands are small, cul-de-sac-shaped orifices lined with a bed of secretory cells that produce a venom of varying toxicity depending on the species. The slimy skin of amphibians is due to the presence of these mucus-secreting cutaneous glands; in Urodeles, these aggregates of multicellular glands give the skin its rough appearance. Implanted in the connective tissue, these glands are covered with cells that form an oval volume with an annular excretory duct protruding from the epidermis. The secretory volume lined with secretory cells is roughly spherical in the case of mucous glands and more oval in the case of granular glands.

Directly connected to the connective tissue supplied by the lymph, the outer layer is made up of smooth muscle fibres which form a ring at the top of the excretory canal. I . the epithelium of the secretory cells is very different if we compare the mucous glands which produce mucin and the large granular cells which secrete toxic substances.

The profusion of these glands is subject to the conditions of the external aquatic or aerial environment, depending on the species of Urodeles, and their development probably depends, in conjunction with the interstitial stem cells available to assemble into aggregates, on the sensitivity of the nerves, the irrigation by lymph and the capillaries situated on the surface of the connective tissue, close to the surface of the epidermis. Their formation is the result of fluctuations in the external environment (temperature, humidity, pressure, luminosity) and the internal configuration we have just described, which cause an adjustment in their number and distribution. In particular, stimuli such as those perceived in the optic lobes for visual sensitivity, which trigger the instinctive contraction of the bucco-pharyngeal system, combined with those felt in the spinal cord in the event of a dioxygen deficit, with excess carbon dioxide and toxins being evacuated to the glands by the skin cells. These stimuli are thought to be conducive to their formation, growth and number, which begins with the establishment of a solid cell bud that starts with nucleation, probably induced by lymphatic flow, to form an initial aggregate of cells that deepen with the appearance of an enclosed glandular cavity, followed by the formation of the excretory duct. The secretion and excretion system acts simultaneously, externally via this annular canal, and internally by diffusion, in contact with the lymph and capillary network by the reaction of the nervous and hormonal systems acting at the level of the glandular nerves stimulated after a light (UV) and/or thermal disturbance, or by pre-adaptation, by reacting to the aggression of a predator by a visual stimulus or direct capture. The tissue of excited nerve fibrils surrounding the tunica muscularis produces rhythmic contractions that alter the volume of the glands. In response to an aggression of a specific intensity, the nervous system triggers the production of mucus and the excretion of mucus, or of excess toxic substances used as venom, onto the surface of the body. The more diffuse internal excretion gives the euprotect a certain immunity to its own toxin, with part of the venom being released into the lymph, diluted in the blood, with a loss of toxicity and eliminated via the urine. On the other hand, a small quantity would be transported in the ovarian tissue to the germ cells and would contribute to the orientation of the egg. These glands are irrigated in direct contact with the lymph and are simultaneously traversed by the capillary network of arteries and veins, which effectively contributes to local detoxification of the skin and lymph, supplemented by the general blood circulation which detoxifies the body (liver). During the revolution of the Urodeles and Anurans, the secretion of a toxic liquid became selectively available to neutralise a predator. Venomous cutaneous mucus[22] is odourless and toxic to lizards and small mammals; in small doses it is a stupefying poison. However, in higher doses, its pathological action becomes convulsive, causing salivation and cessation of breathing. It is advisable not to touch or handle amphibians to avoid injury. We have just seen that these glands interact with the nervous system and the respiratory apparatus in the form of capillaries and lymph, whose major role we have emphasised.

In the event of thermo-hydric stress or attack by a predator, the nerve impulse causes the excretion of mucus to protect itself from dehydration, and in the event of attack, uses the initial dëtoxination, made more selective by the excretion of a venom, to defend itself against an aggressor by using the toxicity of the products sëcrëtës by the skin cells.

By way of comparison with the Crelenteres, the jellyfish absorbs its food and rejects waste through the central opening of the caelenteron where digestive enzymes act, while their stinging tentacles accumulate venom. By filtering seawater, the cells lining this digestive cavity play a direct part in detoxifying the jellyfish's body, while maintaining toxicity in the tentacles it uses to capture and absorb its prey. Whereas in Plathelminthes, depuration takes place after liquid transit through the cells, with waste elimination taking place via a specific channel that forms the first kidneys, these protonephretes have become the mesonephros of the excretory organs of fish and amphibians, which does not rule out ultra-renal excretion via the skin, as has been demonstrated in Urodeles. In the case of the Euproctus, the toxic substance is thought to be a non-alkaloid acid, a hydrated pseudo-

22 Nouveau traité de pathologie, Charles Bouchard and Georges-Henri Roger, 1914.

lecithin which decomposes on contact with water to give alanine[23]formic acid and ethylcarbylamine carbonic acid. Thermo-hydric stress causes the production of a thick layer of protective mucus, while the drying out of the skin leads to the cessation of cutaneous respiration. Under the action of the sun's rays, as the skin dies, it becomes covered in mucin, comparable to the mucilage that protects algae at low tide.

In fish, multicellular mucous glands produce a mucus that covers the scales. The Greenland shark (**Somniosus** *microcephalus*), which swims between 200 and 2000 metres to feed in waters at less than 2°C, has skin cells that secrete trimethylamine oxide, a toxic substance that is of no use against predators, apart from helping to detoxify the skin. Its organism, subjected to very low temperatures and considerable pressure at great depths, maintains a high level of body salinity.

In the yellow-coloured earth newt (**Triturus** *alpestris*), small mucous glands cover the entire surface of the tegument. The large clusters of granular venom glands are located on the sides and top of the body up to the tail, behind the orbit and on the neck, the target organs of predators. When attacked, the venom can be projected a few centimetres away at the sight of a predator or on sudden contact. While the skin of the Alpine crested newt is smooth in the water, during its terrestrial life it takes on a rough appearance, with conical asperites discernible on the surface of the epidermis, as in the Euproctus.

In the Axolotl, in the event of a thermo-hydric gene, the mucous venom is released by a loosening of the annular excretory canal, whereas the toxic venom granule glands act by contraction. This difference in reaction, passive and slow in one case and highly reactive in the other, demonstrates an adaptive defence response via the cutaneous nervous system to the sudden attack of a predator.

The toad is therefore capable of absorbing and eliminating water and dissolved substances through its skin. Compared with the skin of anurans, the toad's skin is thicker, rougher, drier on contact and less vascularised. In air, absorption through the skin is reduced, while toxic secretion is high. In frogs, the skin is thinner, remains moist and highly vascularised, absorption is faster, and toxicity varies from species to species depending on their adaptability.

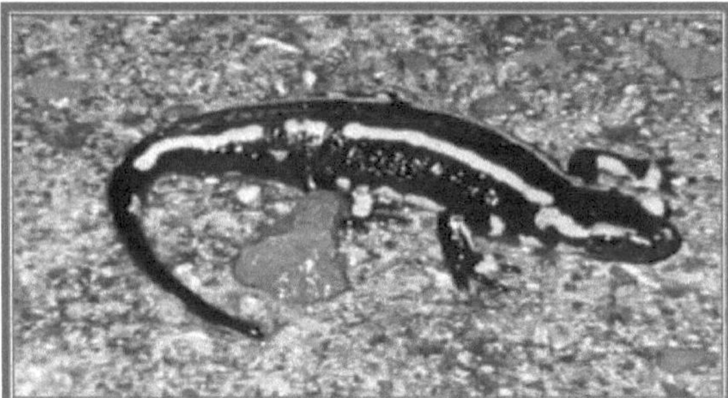

36. Terrestrial salamander, Salamandra salamandra.

For all these species, as the temperature rises, the permeability of the skin increases and

---

23 Alanine or alpha-amino-propionic acid is an amino acid found in Coenzyme A, which is active in the transfer of groups by participating in the oxidation of fatty acids and the synthesis of alanine. An alanine-specific tRNA contributes to protein formation.

the metabolism becomes faster under the effect of thermal and hydric stress, until dryness causes dehydration, followed by irreversible damage to tissues and cells if the amphibian cannot find water or a humid refuge to rehydrate. The roughness of the surface of the skin, its constitution and its thickness do indeed vary according to altitude, light and temperature, moving from a hydrous to a terrestrial environment.

Gaseous exchanges between the water and the skin enable dioxygen to be transferred directly to the epidermal cells and their nuclei, which are highly polarised depending on the depth of the cell layer, through simple diffusion and osmosis across the epidermal cell membranes, by chemical dissociation of the water, all the way to the dermis, whose connective tissue is abundantly vascularised, in addition to the gills, which are very active during the aquatic larval phase, prior to metamorphosis.

Before going into more detail in this chapter, to summarise, respiration through the skin directly feeds the connective tissue irrigated by lymph, which is connected to the capillary networks of the arteries and veins of the general blood circulation. The highly vascularised and innervated connective tissue is coupled locally to the pigments located between the dermis and the epidermis, which take part in the cutaneous respiratory process. This highly innervated unit is sensitive to internal physiological conditions and highly reactive to environmental fluctuations. This respiratory system takes part in the capture, diffusion and fixation of dioxygen in the skin cells, and the lymph of the connective tissue transports it in the opposite direction to the general circulation, which is supplied by the blood, with a complementary supply of dioxygen from the bucco-pharyngeal respiratory system, which is activated in water and even more so in the open air.

When the Euproctus is in a state of respiratory stress compensated by the bucco-pharyngeal intake, if this situation of discomfort persists, the overcompensation further stimulates the primitive pulmonary respiratory system, still reduced to its simplest expression.

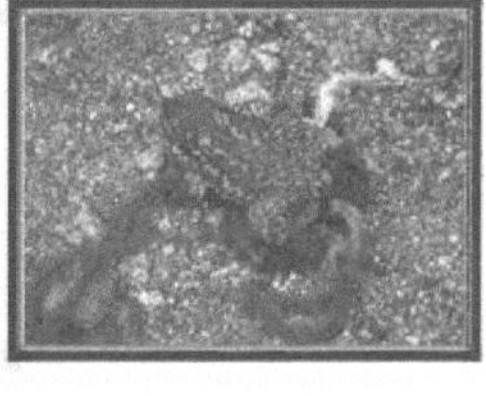

*37 and 38. Common toad, Bufo bufo and half devoured toad,
under the repulsive effect of the venom.*

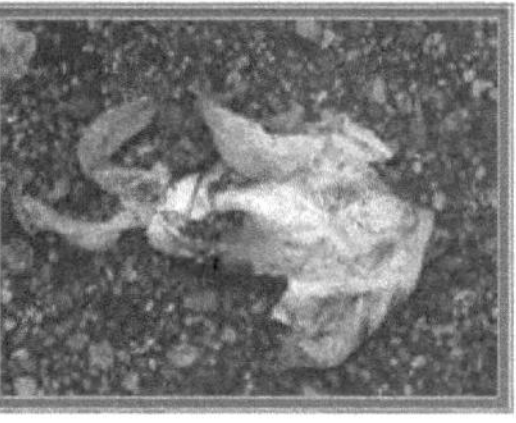

39 and 40. Red frog and skinned frog,
completely devoured (Mont Lozere, 1,400 metres)

If we adopt this evolutionary hypothesis, among others, the shortcut to breathing through the skin leads to efficient evacuation of cellular waste from the epidermis into the dermis, which is highly vascularised by lymph, towards the mucus-secreting and excreting glands, which then proceed to directly detoxify the cells and connective tissue by producing a type of gel through specialised glands. This more or less thick mucus proves very useful when it is released into the open air, and as a result of its acquired toxicity, it has become selectively positive for the survival of Euproctes, warding off potential predators. In the same way, by momentarily avoiding lethal dehydration, it gradually constituted a serious selective advantage during its first terrestrial movements. Many amphibians secrete toxins from specialised granular glands in a liquid thicker than mucus. Depending on the species, these secretions include amines, polypeptides and alkaloids such as batrachotoxin and tetrodotoxin. They are highly effective in adults, causing burns and numbness that force predators to drop their prey, and sometimes lead to fatal poisoning, which is why they are used by certain hunting tribes to impregnate their arrows. These toxins are the subject of pharmaceutical studies with a view to using them in infinite doses in medicines.

In the air, depending on the outside temperature, light and humidity, mucus rapidly accumulates on the skin, providing protection against dehydration and playing a role in gas exchange by modulating pressure differences between the air and the internal environment. Detoxination can then be amplified, as in certain more terrestrial newts and toads, whose more numerous and highly toxic mucus glands are clearly differentiated. According to our approach, based on our observations and inspired by the data of R. Despax, cutaneous respiration in water gives autonomy to the peripheral cellular metabolism constituted by the surface of the skin. The epidermis and dermis act as a membrane between two practically similar external and internal hydric environments. As a result of the chemical dissociation of water, exchanges of dioxygen, hydrogen (proton and electron), electrons, ions, trace elements[24]The exchange of oxygen, hydrogen (proton and electron), electrons, ions, trace elements and iron, the active constituent of haemoglobin[25] (appendix 5), which is ubiquitous in high-altitude pools, reaches the skin cells directly. By extension, this localised cellular respiration enables oxygenation of the connective tissue, which is highly irrigated by lymph in peripheral blood vessels, initially in the opposite direction to the general circulatory system. The latter, which will become more elaborate

---

24 A vitamin C deficiency leads to skin disorders.
25 Oxygen dissolves directly in the cells of the epidermis and in the plasma of the blood that irrigates the vascularised dermis, where it binds to haemoglobin, which carries four molecules of oxygen.

in mammals, we can see the beginnings of this derivation towards the general circulation by observing bucco-pharyngeal respiration and the still underdeveloped lungs that supplement cutaneous respiration in many Urodeles. In water, external pressure plays a part, by simple diffusion and osmosis in the epidermal cell layers, in balancing the external hydrous environment and the internal cellular environment, where the thickness of the cell layers, the roughness and the constitution of the cellular aggregates vary according to altitude and climate, in a hydrous or aerial environment (appendix 4). Air contains more oxygen than water, and in an oxygen-poor aquatic environment, the euproct seeks out areas of current turbulence to improve ventilation of the skin cells, as carbon dioxide tension remains low in these conditions.

However, in the open air, which is richer in oxygen, cutaneous ventilation is less than in water, and carbon dioxide tension increases, which in the event of sustained effort requires respiratory compensation directed primarily towards the bucco-pharyngeal mode, extended by incidental respiratory overcompensation to the alveoli in demand. After a long period of adaptation through genetic variation and natural selection, these saccules will gradually be adapted into functional open-air lungs. Changes in carbon dioxide tension modulate the acid-base equilibrium in aquatic and aerial environments in different ways, through fluctuations in the concentration of hydrogen ions. The hydrogen potential *(pH)* corresponds to the gradient of *H+* protons, and these variations in acidity play a part in enzymatic activity, which at an optimum *pH* gives maximum activity. The organism's *pH* remains almost identical between water and air, and this buffering effect results in an increase in carbonate ions offset by an increase in chloride ions (*Cl-*), with a notable difference in the concentration of (*Cl-*) between the aquatic and aerial environments. This is particularly true of amphibians, which pass from a larval state to a more or less terrestrial adult state through the process of metamorphosis. During hibernation, dissolved sodium chloride, combined with antifreeze glycopeptides, helps to lower the freezing point of cells. When the metabolism slows down, the reduction in heartbeat and the contractions of the lymphatic vessels lead to a drop in dioxygen and an increase in carbon dioxide in the blood, the latter creating an acidosis that causes the body to fall asleep. The increase in carbonates is balanced by chloride ions. Under the action of cold, concentrated sodium chloride salts accelerate the effectiveness of anti-freeze glycoproteins, which replace water in skin cells, preventing them from freezing. In the mitochondria of these cells, acidosis causes an additional supply of reactive protons *(H+)** in the intermembrane space, which amplifies the mitochondrial threshold pulse field *El'(ll+)** and the production of ATP at low temperature. This excess of energy *A€AV(H+)** is manifested by thermogenesis, which maintains a minimum of vital coherence at around 0°C, corresponding to the reduction in pulses from the main short systolic creur, which is supplemented by the lymphatic creurs with long contractions, spread over five consecutive cycles. In the absence of a more complete exploration of this process in the Euproctus, I am merely opening up the field of research whose discrete effects I mentioned in my essay on low biological energies.

Incidental acidosis is effectively compensated for by the adjustment variable of increasing carbonate ions, which may have occurred between the Devonian, Carboniferous and Permian periods. The increase in oxygen tension in the air led to a rise in carbonate concentration and its re-equilibration when the external temperature rose, an important adjustment variable in the enzymatic activity of cells as a function of acidity. During periods of natural climatic warming, the heat caused an excess of energy expenditure by the metabolism, and thermal stress set in, putting pressure on the body to produce a protective and excretory mucus, which detoxifies the resulting acidity. This stress acted on the activity of the bucco-pharyngeal respiratory tract, which was still not very efficient, but already existed and was selectively sufficient to replace cutaneous respiration. In addition, it triggered the search for water or hydrated food when moving to a damp, sometimes dry and cold terrestrial environment, which meant that amphibians underwent a real physiological transition between these two environments. During their evolution, they benefited from the process of metamorphosis with gills, with the respiratory properties of the skin providing a respiratory bridge to the oral-pharyngeal pathway, and to a lesser extent to the developing pulmonary pathway. The passage of more or less dissociated water through cell membranes is not neutral, particularly for charged particles such as *H+* which determine the acidity or basicity of the biological terrain, as well as for the electrons and ions involved in metabolism and cellular respiration, particularly iron. Their passage through the cell membranes and organelles such as the mitochondria determines the free potential energy delivered in the form of *ATP,* depending on the supply of *ADP* and phosphate *(P) to the* skin cells. We have described the epidermis, which is made up of two to three layers that establish a thin differential thickness between the cells that are joined together and superimposed, and which, at the level of the cell membranes, engage in transport that will satisfy the connective tissue's need for dioxygen.

This arrangement contributes to the structural stability and vital dynamics of the organism, which is kept in an unstable equilibrium around a viable average by the food ingested by the Euproctes, by contributing to the basic metabolism that feeds the organs in order to supply the general blood circuit in addition to the localized and autonomous blood circuit of the skin cells. It is made up of a thin layer of active cells arranged in layers comparable to more or less liquid crystals which benefit directly from the dissociation capacity of water, the products of which can be directly recovered by cells with polarised nuclei. According to our hypothesis, skin cells breathe in 'pulses', absorbing dioxygen and releasing carbon dioxide, the diffusion of dioxygen taking place gradually as far as the connective tissue, which is bathed in lymph. The latter spreads out in opposition to and in conjunction with the blood capillaries of the general blood circulation, without discharging waste products on a long journey through the body, but by direct, localised detoxification directed mainly towards the glands included in the skin. Membrane transport in epidermal cells allows ions and molecules to pass through the membranes from the outside to the inside, thanks to the physical and biochemical properties of the membranes, which are made up of lipids and proteins embedded in the membranes.

These basic protein complexes are very involved in the transport of oxygen molecules, ions and electrons, as well as the translocation of protons in the mitochondria. In order to generate these flows in the cells, there are transmembrane gradients that affect the epidermal cells with action potentials of a few millivolts that provoke these pulsating reactions. These gradients are crucial to cellular activity and are initiated by the interaction of electron flows during oxidation-reduction, in correlation with the electromagnetic fields of the highly reactive protons in the mitochondrial intermembrane space, which act instantaneously to modify the shape of the atoms and molecules in the cellular environment. On the basis of the following principle, the quantum field pulsed at saturation of "protons and photons", at threshold, would reach a higher celeration than the biochemical reactions. As a result, this quantum field and its derivatives would constitute the coordinating unit of the cells, which would instantaneously disrupt the chemical conformations of the atoms and molecules in the cellular environment, modifying the biochemical interlocking, while at the same time triggering thermo-hydro-dynamic fluctuations and dissipations. This hypothetical process would contribute to passive and/or active transport by spontaneously 'feeding' the cell with metabolites during the pulsing phase of the skin cells. This short activity would precede the long intramitochondrial metabolic phase of the Krebs cycle.

In the Euproctus, cutaneous respiration coupled with excretion, autonomous and localised, conditions its development, metamorphosis and, consequently, the regeneration of its organs. This sensitivity is remarkable in the skin, eyes and intestines, which are capable of simultaneously undergoing destructive histolysis and structuring histogenesis. For oxygen dissolved in water, its transport across membranes can be considered passive, insofar as the external environment is considered to be 'almost' similar to the internal environment, with the slight water overpressure depending on altitude and water depth. Oxygen molecules diffuse across the cell's membranes, the differential osmotic pressure generating this movement; for water, this simple diffusion is facilitated by aquaporins, selective channels of around 2.8 Amstrong for the passage of a single molecule of water and its inter-membrane transport, impermeable to protons. As we have just described, in the mitochondria, protons are trans-localised by specific transporters which supply the cell with energy in the form of ATP. At the same time, these active transmembrane transports are carried out by specialised protein complexes assigned to the transport of ions, electrons and protons. The potential of a membrane varies on average between -50 and 200 mV, depending on the permeability of the cells that make up the skin tissue, whose sensitivity to light, temperature and oxygenation we have mentioned in the Euproctus. The organism's need for oxygen, hydrogen, electrons and ions is dependent on the thickness of the skin, the constitution of which is linked to the altitude and temperature of the hydric environment, which provokes a punctual, non-definitive adaptation that, depending on the circumstances, is exposed to natural selection through these reversible bio-ecological modifications or through irreversible morphogenetic variations. This long-term process initiated by genetic and epi-genetic devices is both conservative and, in the long term, initiates innovative adaptativities:

"These evolutionary contingencies of a stochastic nature are manifested by a structuring and/or de-structuring adaptativeness that irreversibly distorts species through their morphogenetic variations, which are fundamentally random in their relationship to the chance circumstances of natural selection, from generation to generation".

This is the fundamental paradigm of complex evolutionary processes.

To understand these limits to reversibility, let's look at the dynamics of the mucus, epidermis and dermis complex. The formation of an adsorption film that accumulates on the surface of an adsorbing condensed phase is formed by the skin and the mediating mucus. This thin composite layer is physically adjusted to its environment by a limiting surface concentration that forms a dense monomolecular thickness of the adsorbed substance, which may be either directly the water of the external and internal aquatic environment, or indirectly the humidity of the air. As a result of this adjustment variable, this two-dimensional film is modified, and due to this induced metastability, the mucus adopts an optimised configuration of its biological milieu, to the environmental conditions. In the case of the skin, and more specifically the epidermis with its mucus layer in contact with saturated steam, there is a dependence on a chemical potential, or more precisely a biochemical potential, because the mucus film is loaded with minerals (pigments, ions) and other biological components (toxins, mucin), in its ratio to that of the chemical potential of the water and absorbed moisture, which defines two conditions of equilibrium:

- In water, the very thin film is virtually indistinguishable from the surrounding water mass, optimising local oxygen uptake and detoxification.

- In air, depending on temperature and humidity, the film remains stable for a thickness below a limit corresponding to its equilibrium with the saturating vapour of the humid environment, then from this limit point up to a maximum, it is distinguished by a variable metastability that determines the secretion and thickness of the excreted mucus. The glandular system of the skin, which is very soLLIcitë to maintain it, adapts by means of an internal nucleation initiated by the circulation of lymph on germ-ions with the structuring effects mentioned earlier, followed by the formation of cellular aggregates which form mucus glands. It is the fluctuations and dissipations of this ultra-sensitive, metastable mucus that help shape the active cells of the epidermis and the innervated, doubly vascularised connective tissue of the dermis. Their epigenetic reactions to external disturbances will react significantly to adaptive variations in their internal physicobio-chemical potential, through the development of additional glandular clusters, some of which will prove toxic. In the open air, in the absence of water, depending on the ambient humidity and the dehydration caused by the outside temperature, metastable mucus plays its role as an intermediary for the respiration of skin cells. It temporarily provides the transition between the liquid and gaseous environments to which Euproctes are exposed during their journeys out of the water and during their terrestrial life, mainly during hibernation and more occasionally in the larval stage. The composite skin, the adsorbent mucus, a superficial film of absorption, goes through this short phase of metastability, which we believe plays an adaptive role in response to variations in altitude, temperature and humidity. As we have just shown, the changes in external state are linked to the interdependence of the physico- biochemical potentials of the stratified skin-mucus components, which interact to maintain internal equilibrium by adapting the appearance of the rugosites made up of cellular aggregates, while increasing their number and activity, to the point of differentiating the cutaneous glands. This state is very unstable in the open air compared with the aquatic environment, where oxygen is thirty times more concentrated and diffuses more or less rapidly across the surface of the skin, depending on the diffusion system specific to each amphibian family. In most Urodeles and Anurans, we find the mucus, epidermis and highly vascularised dermis model provided with pigments, thermo-photosensitive sensors in addition to those used to transport dioxygen, such as haemoglobin and oxyhemoglobin from the general circulation. From a certain stage of evolution in amphibians and urodeles, this adaptive sensitivity was taken over from the cellular pulsations of cutaneous respiration by buccotracheal respiration, which is very rhythmic in anuran amphibians. Gradually, the pulmonic respiratory system became much more complex, with pulmonary alveoli that suggested the use of lungs in mammals, favouring the general circulation of blood through the creur at the expense of cutaneous respiration. The concentration of oxygen in air is 209 ml/l, whereas in water it is only 7 ml/l, giving a ratio of 1/30. The transition from the aquatic to the terrestrial environment posed and continues to pose a number of problems when :

-    As drought worsens and humidity levels fall, dehydration sets in, causing water and heat stress that leads to short-lived compensations, and possibly salutary over-compensations, followed by aggravating pathologies and death.

-    On land, we need to consider the accentuated effect of gravity (appendix 4) and the loss of lift due to water, which increase energy expenditure during movement. In response to these constraints, a process of water retention took place through anatomical and physiological adaptations of the blood, kidney, brain and hormonal systems, involving a necessary metamorphosis.

Thus, plasma proteins are more numerous in terrestrial amphibians than in those that remain aquatic, and vasotocin slows down diuresis by participating in osmotic balance. The organism's response to thermo-hydric stress in order to survive cold or heat takes place in ëvolutive stages that range from the gradual installation of various metamorphosis processes to the practice of hibernation. First comes an alarm reaction, which manifests itself in a few minutes as the organism adapts to the thermal aggression accompanied by a movement to take shelter or drink immediately, if not by causing a migration, to a less hostile place. The endocrine glands release hormones that speed up the heart and respiratory rate, increasing blood sugar and sweating, and in this case skin secretion. The pupils dilate, intensifying the role of the optical system, while digestion and metabolism slow down. The body's resistance kicks in, regulating the disturbances via heat shock proteins until they can no longer compensate for them. This leads to a period of exhaustion, with irreversible effects on the body. This neurobiological response at the level of the cells and the organism is prolonged by survival behaviour, and possibly by ultimate overcompensation which manifests itself in ancestral reactions and sometimes innovative behaviours with social, cooperative or competitive consequences that are not insignificant for the social instincts of animals, and which induced in the human species a unique civilisational reversal effect, discussed in the final chapters.

For Lissamphibians, the challenge has been vital to ensure the transition between the aquatic and terrestrial environments, wet to dry, as we have mentioned that gas exchange takes longer to pass through the skin in the 'dry' state. In the course of their evolution, this meant that skin structures had to be kept moist within certain powerfully adaptive stress limits, during the processes of metamorphosis and hibernation.

The amphibian organism underwent adjustments to the changing environment by favouring pre-adaptations such as bucco-pharyngée respiration. Associated with the contraction of the optic lobes, the rhythmically moving buccal floor had connective tissue that was previously sensitive to the supply and forced diffusion of dioxygen by these contractions, while initiating pulmonary respiration, reduced to a few saccules, a function that gradually expanded. We will look for the origins of these respiratory modes in the phylogeny of Lissamphibians and by observing their ontogeny, as well as describing reversible adaptations to altitude-dependent living conditions in the Andes and to those experienced in space, out of gravity (Appendix 4.), compared with our observations of Euproctes in the Hautes-Pyrenees. During hibernation, its metabolism is reduced to a minimum, in a state of torpor, it is immobile, without movement and without feeding, breathing exclusively through the skin as we have just demonstrated, relayed by the lymphatic creurs. ***Salamandrella*** *keyserlingii,* which lives in northern Siberia, loses water to increase its salinity and lowers its freezing point using antifreeze (glucose, glycerol). The hibernation period contributes to the development of sex cells in the field of low biological energies, a field of research that remains to be explored... After describing the respiratory processes of the Euproctus, and noting the adaptive capacities of their respiratory systems, which are thought to be autonomous at the level of skin cells, in order to correct our models of their evolution, we need to assimilate other data relating to ontogeny, metamorphosis and organ regeneration. Let's start by discussing the initial conditions of embryonic development in the Euproctus.

42. An amphibian hibernating in a stream
on Mont Lozere, 1,400 metres.

Chapter V

Ontogeny of the Euproct.

The evolution of the Urodeles, proceeded from a period of differentiation between the Anurans and the Urodeles, particularly with regard to the respiratory, cutaneous, oral-pharyngeal and pulmonary systems. The Urodeles diverged without major modifications into several species of salamanders and newts during a stabilising evolution that led to the Pyrenean Euproctus, while retaining respiration and cutaneous excretion. The ontogeny of **Calotriton** asper asper, and more specifically the modelling of its embryogenesis, requires in-depth knowledge of the stages of embryonic development in Anurans and Urodeles. From fertilisation of the egg by a spermatozoon, its development takes place in four stages: segmentation, gastrulation, neurulation and other stages of the caudal bud. To develop a model of the initial conditions for the development of the Euprotect, we used data from a practical manual of experimental embryology to simulate *the segmentation phase,* updated by a more specific study of the Axolotl egg proposed by J. Signoret and J. Lefresne. Research by biologist Francois Gasser[26] at the biological station at Lac d'Oredon, a few kilometres from the Moudang barns, focused on observations of the initial stages of development of **Calotriton** asper asper. This researcher experimented in the laboratory with rent's maintained at a temperature of 12°C, noting a high mortality rate due to the conditions of the experiment in crystallising tanks, with the fecund egg requiring high oxygenation during its development. During our observations, we recorded the natural conditions of water agitation and the critical temperature threshold as a function of daily sunshine. We will discuss the various external factors that affect the reproductive stages and fecundation of the egg in the Euproctes streams around the Moudang barns. Let's take up and comment on F. Gasser's methodical description of the developmental phases that he described in September 1963:

- The non-segmented fecund egg is laid under stones, sheltered from excessive water flow and protected from predators. It is described as large and yellowish-white in colour, with an average diameter of three millimetres, spherical in shape, flattened at the animal pole and with barely visible pigmentation. Certainly sensitive to the sun's rays, attenuated by the height of the water, the egg is covered by a thin, rigid and opaque shell, which protects the inside, which is made of an embossed translucent gangue.

- After careful degangulation, F. Gasser observed the egg and decided that segmentation, gastrulation and neurulation corresponded to the chronology of morphological stages described by Gallien and Durocher (1957) for **Triturus** *helveticus,* which serves as a reference for the model below.

Segmentation lasts four to five days, gastrulation six to nine days and neurulation nine to fifteen days. We will return briefly to the caudal bud and larval stages with gills, and dwell at greater length on the initial phases of segmentation of the egg activated by a spermatozoon.

Good conditions of water oxygenation at an optimum temperature of 17°C, protected from light, are essential to initiate egg division when the mature oocyte's sensitivity is triggered by an activating spermatozoon. This early stage of embryonic development turns out to be

---

26 Les bases embryologiques et genetiques de revolution de l'reil : conservatisme et innovation, Francois Gasser, Pour Darwin, puf,1997.

a discrete process of transition from linear to non-linear, when cell divisions diverge, becoming highly asymmetrical with an asynchronous tendency between the vegetative and animal poles. Cell division reveals a rate of divergence that is sufficiently structuring, because it is genetically and epigenetically contained, maintaining the coherence of its components to give rise to a viable form of embryo. Understanding the initial causes of the asymmetry of the egg seemed to me to be fundamental, and to understand them it was necessary to detail the first stages of segmentation at several levels of resolution, from atoms to molecules, from proteins to the cells that make up the dividing egg of a potential animal.

42. Mating of Euproctes.

It is clear from F. Gasser's experiments that the development of the Euproctus is slow, with the embryonic phase lasting forty-five days before the larva hatches, and late feeding only beginning after three months. This stage of larval development seems to me to seriously involve exchanges between skin cells and the hydrous environment, during cutaneous respiration, a theme developed at length in the previous chapter.

Like Despax, Steiner and Stoll, F. Gasser drew our attention to the presence of horned claws at the tips of the fingers of the larva's legs, which proved useful for moving around and staying underwater on the rocky bottom during this larval period, and during its terrestrial outings, which we actually observed. Sexual differentiation between males and females occurs on average at sixty-five minutes. These authors also describe :

"the most common forms on the skin, formed by cellular aggregates".

A characteristic of cutaneous sensitivity that we have retained as an adjustment variable for the adaptativeness coefficient of the Euproctus, in relation to altitude, temperature, humidity, water pH, oxygenation and illumination of the aquatic environment, parameters used in the models and simulations. In the larval stage, the dark-coloured, short, foliaceous gill body remains subject to oxygenation conditions, and we have already mentioned the importance of oxygen mixing in streams. According to F. Gasser, slow development, the causes of which we discuss at length, is likely to cause a delay in differentiation between the animal and vegetative parts of the egg, and is therefore likely to have an effect on the speed of segmentation, which would lead to the desynchronisation of cell divisions and the loss of symmetry between the vegetative and animal poles. The divergent divisions of micromeres and macromeres had to be taken into account in our experimental approach, and the underlying causes investigated.

Simone Rouy, has shown that this Urodele has a very progressive larval development and a very slow growth after metamorphosis, the two external factors that frequently emerge from their observations are always **temperature** and **oxygenation**, to which we add **luminosity and ultraviolet rays, the pH of the water and the presence of iron, and to a certain extent gravity and terrestrial magnetic fields,** factors to be weighed in the water and at altitude. These external factors are linked to the internal factors by choosing free energy as the main parameter:

"The segmentation of the active a'iif can be evaluated in relation to a coefficient of adaptability representative of its divergence, in equivalence with the minimum free energy necessary and sufficient to cause its division".

We have highlighted the major role played by these parameters in cutaneous respiration and the formation of skin containing light-sensitive pigments, while spermatogenesis and sperm evacuation take place under strict temperature conditions.

Spermatocytes form at temperatures above 12°C, below which spermatogenesis is blocked and the spermatozoa degenerate, requiring high temperatures of between 21°C and 25°C for their evacuation. Most eggs contain melanin (amino-sulphur); in water and under the summer sun, this pigment increases their temperature while protecting the egg from the penetrating power of ultra-violet "A" rays with a wavelength of between 400 and 315 nanometres and an energy of 3.10 to 3.94 electron volts. This radiation can cause mutations and activate free radicals, but it can also increase the amount of protective melanin in the egg.

After these preliminary observations, we will deal with the division of the egg under its triple aspect, quantum, biochemical and thermo-hydro-dynamic, by considering its volume for an average diameter of three millimetres. The potential energy of the components of the animal pole and the vegetative pole of the oocyte are subject to fluctuations-dissipations during its *"energising activation"* by a spermatozoon, whose tail has a sheath lined with mitochondria, as does the acrosome of the head. In the fëcondë egg, the perturbation induced by the spermatozoon causes its internal pressure to vary within a constant volume, activating a minimum ëfree energy that would be the initial condition for the transition to non-linearity at the origin of the gënëtically controlled divergences during the segmentation phase. The scientists Monique Clergue-Gazeau and Jean Claude Beetschen (1966) provide us with information on the reproduction of Euproctes observed at two sites located at very different altitudes.

At 2,328 metres in the Hautes-Pyrenees, the adults seek out the oxygenated waters of torrents to mate; this amplexus causes the cloacal lips to come together during fecundation. After the long phase of reproduction by entanglement, the females head for calmer areas with sufficient oxygen to deposit the second instar larvae in isolation, at a shallow depth. The larvae hatch and remain in these calmer waters, before moving on to agitated areas and, as they grow to adulthood, venturing into streams with turbulent waters. Their comments state that during metamorphosis, in good weather, the larvae move towards water-soaked plants and mosses, which is what we have actually observed; in bad weather, they bury themselves in the mud, and they conclude by saying that metamorphosis requires fairly strict conditions.

In the Pyrenees ariegeoises at 850 metres, the observation took place in a pool where they were interested in their return to the water, which does not seem to be directly linked to reproduction, but certainly to more physiological and metabolic causes, such as predominant skin respiration, directly linked to climatic fluctuations and the fairly severe meteorology in the mountains. The search for food is more accessible in the water during the larval period with gills and in the adult state by settling in small, calmer retention basins to capture insects that fall into the water.

If these Euproctes lead similar lives in almost identical environments, it is the climatic variations closely linked to altitude and season that modify their way of life and reproduction. We will see in the course of the modelling that the physical parameters of their biotopes, which are very closely correlated with their internal biochemical and physiological factors in an aquatic environment, need to be carefully assessed. In particular, during the segmentation phase, in addition to temperature and humidity, we need to take into account, depending on altitude, in the water, solar infra-red and ultra-violet radiation, the effects of gravity and low turbulent hydrostatic pressure, which act as a kind of microgravity, on the polarisation, cohesion and adhesion forces of the cells of the fecund egg. This compares with the experiments carried out in space over the last few years, outside gravity, the consequences of which on the morphogenetic processes of embryogenesis are beginning to be understood.

Before going on in detail about the chronology of fecundation and the first stage of activation of egg development during segmentations, I would like to go back over the main points.

27 In the absence of euproct eggs, segmentation observations had to be completed on frog eggs from Mont Lozere.

hypotheses previously formulated on the coordination of cellular activity by a quantum field of protons and photons at saturation, pulsed at threshold, in the intra-membrane space of the mitochondria. This unitary field would be sufficiently effective to coordinate the brief phase of cellular and intercellular activity according to the density of the mitochondrial network, contributing to cell cohesion and enabling the animal to breathe and feed to ensure its vital coherence. In our experiment on low biological energies, we suggested 'cold' quantum solutions associated with a possible bioplasma, a source of vortices, probably at the origin of nucleation, and 'hot' thermo-hydro-dynamic solutions during structuring fluctuations-dissipations. This proton field, which is pulsed at the threshold, is said to contain waves with divergent derivatives in space-time, which are discreet to say the least, and are manifested at long distances. Could they be involved in ontogeny?

If we consider an initial activation wave caused by the spermatozoon during fecundation, after fusion of the acrosome, there would be "***an induced field, the divergent derivatives of which would spread out in space-time***". This gives full meaning to the notion of a morphogenetic field, the free energy of which would cause non-linear divisions of the egg, under the control of the genetic system interdependent on this supply of energy. Without reducing the cell to the atomic level alone, I insist that there is a reciprocal loop between the protein complexes of the inner membrane of the mitochondria which supply free energy in the respective forms of fields (electrons, protons) to produce ATP, in a co-action with the genes of the mitochondrial DNA and nuclei which are expressed in the dividing cells of the animal. This mosaic of interactions, based on the concept of unitary levels of integration, goes well beyond a simple hierarchy from gene to trait, insofar as we consider that the coordination of cellular activity by unitary fields gives full justification to Emergence, of proteins, of cells, and of the animal, according to the biochemist Faustino Cordon.

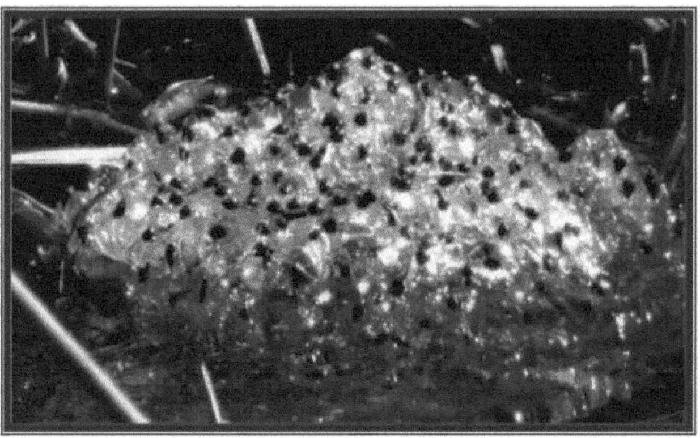

43. Cluster of red frog eggs
(Mont Lozere, 1,400 metres).

The inner membrane of a mitochondrion is made up of four protein complexes that transport electrons. This flow is concomitant with the translocation of protons from the alkaline matrix to the more acidic inter-membrane space by three protein complexes, *I, III and IV*. This proton-motive force is involved in the synthesis of *ATP* by oxidative phosphorylation of *ADP*, according to Peter Mitchell's chemo-osmotic theory, and protons are introduced into the matrix by complex *V, ATPase*. On its "way [27]" the proton phosphorylates Adënosine Diphosphate *(adenine-ribose-P~P)* into Adënosine Triphosphate *(adenine-ribose - P~P~P)* on the ribose, the chain bond which continues with the oxygen-bound phosphates is rich in ëenergy *(P~O)*. Furthermore, the mitochondria concentrate calcium, Ca^{++} acts on the permëabilitë of this membrane via Permëabilitë Transition Pores *(PTP)*, their opening would correspond to an excess of calcium following oxidative stress.[28]. Oxygen deficiency, by reducing the ënergëtic *(P~O)* ratio, causes these channels to open as a result of an excess of calcium, which dissipates the membrane potential and critically alters the ionic balance (electrons, protons, ions), causing the leakage of essential metabolites. When the permëabilitë of the membrane is increased, the inter-membrane *pH is* modiim, and we presume that the field of protons and pulse photons, which is no longer saturated, does not reach a sufficient interaction threshold with the atoms and molecules of the cell, to trigger its activation and the coordination of their rearrangement, which is necessary for the short phase that supplies the cell's cytoplasm with metabolites, carbohydrates, amino acids and fatty acids destined to form acëtyl-coenzyme A. Its supply guarantees the long intramitochondrial mëtabolic

---

27 By a probable tunnel effect, the proton, both wave and particle, is translocated to ATPase receptor sites in a veritable biological motor measuring a few hundred Armstrong ($100*10^{-8}$ cm).
28 Free radicals induce dëlëtëres chain reactions, such as lipid peroxidation and the cleavage of protëines, and cause DNA damage.

phase of the Krebs cycle, leading to the supply of CO_2, but above all to the supply of electrons and protons to the prokine complexes involved in their transport in order to maintain the proton gradient that initiates the field that conditions the coordination of cellular activity. To what extent is it involved in fëcondation and development?

In such a complex process, this hypothesis, which I support in my proposals, does not exclude other causaliks, Jean Claude Beetschen[29] has written preëcisëment on the development of Amphibians, he mentions a megalocal distribution of the yolk in the hëtërolëcithe egg, he observed the following initial conditions:

- Less cytoplasm in the Hemisphere of the vëgëtative pole compared with that of the animal pole.

- While the vitelline platelet gradient is denser, they become larger from the animal pole towards the vitellus-saturated vëgëtative pole.

In amphibians, the yolk constitutes a potential energy reserve accumulated in the oocyte in the form of platelets containing phosvitin, a phosphosërin-rich prokine capable of binding *Fe3* cations$^+$, iron is very present in Euproctes ponds, Ca^{2+} and $Mg2^+$, and lipovitellin, a lipophosphoprokine active during the transition from one segmentation phase to another. Phosphoric acid plays an important role in the maintenance of cytoplasmic *pH*, which conditions gënëtic conformik, as well as the release of protons and electrons engagedës in respiratory processes at the level of prokine complexes in the inter-membrane space of mitochondria.

The mitochondria gather around the centriole, the vitellus nucleus of the vëgëtative pole of the oocyte. These organelles play a major role in fëcondation and egg development and are involved in the synthesis of iron-sulphur protëins and nucleotides.

Phosvitin, whose precursor is vitellogenin produced by hepatic cells, is transmitted to the ovary by the blood.[30] It is a phosphoglycoprotein composed of aggregates of polypeptides, a-phosvitin with 3% phosphorus and в-phosvitin with 10% phosphorus, with a high phosphoserine content. Its linear conformation at a neutral *pH* takes on a more compact appearance at a *pH* below 2. Phosvitine is a metal transporter that binds iron until saturation. In the event of long-term acidification to *pH 6*, ferric ions induce proton dissociation and the formation of a stable iron-phosvitin complex. By taking up Fe^2 it reduces the production of hydroxyl (*OH*) free radicals, acting as an antioxidant. However, after a lack of oxygen, re-oxygenation leads to a supply of superoxide (*O* and $H2O2$) associated with F^2 , giving *OH'* which, by hydrolysis, would lead to tissue damage. Is this process involved at metamorphosis in the relationship between histolysis and hytogenesis? We have observed Euproctes larvae with gills living in iron-saturated tanks.[31] their gill and skin respiration would be facilitated.

During vitellogenesis, lipids are stored in the fat bodies and mobilised in the liver to be progressively transformed into vitellus in the oocytes. The ovaries contain small non-vitellogenic oocytes and larger ones rich in vitellus when they reach maturity. In contrast to the vitellus in the vegetative pole, in the animal pole, the nucleic acids are distributed in

---

29 Jean Claude Beetschen, Honorary Professor at the Paul Sabatier University in Toulouse, has provided us with documents and references on the study of Euproctes in the Pyrenees.
30 The toxic venom of the skin glands, diluted in the blood, acts on the maturation of the ovaries.
31 At the ferruginous spring (1760 metres), from July to September, the pH is between 7 and 8 with an alkaline tendency, and the oxygen content is 92 to 98% in running water at a temperature of 10.5°C. Aquatic plants thrive here.

a gradient, abundant in proteins, ribosomes, rRNA and mRNA, all of which are called upon during the activation of the second egg, from the first divisions onwards, directing the embryonic development of the larva. Could this difference in the distribution of excess vitellus at the vegetative pole compared with the distribution of cytoplasm at the animal pole be at the origin of the dissymmetrical segmentation accentuated by the fluctuations and dissipations initially induced by activation of the spermatozoon during fertilisation of the egg?

Although the mitochondria of the spermatozoa are involved in their movement to the oocyte, it is mainly those of the oocyte dispersed in the cytoplasm that are redistributed in the two pro-nuclei after fertilisation to provide energy in the form of ATP. This activation phase is highly dynamic, with the cytoplasm localised at the periphery of the animal pole, which undergoes major reshuffling after conditional activation by the spermatozoon. The movement of the sperm towards the centre of the egg, causing depigmentation around the equator, results in the formation of a grey crescent, the causes of which are multifactorial, difficult to detect and differentiate, and which we are attempting to explain using experimental simulations. Although the central mass appears immobile, in the continuity of the spermatozoon's path, the cross-section of the egg taken by the experimenter shows a clear pigment trail indicating superficial movement of the cytoplasm on the surface of the egg, with the formation of this grey crescent opposite the point of entry of the spermatozoon, a structure that will play a role in morphogenesis.

This initial condition for obtaining relative symmetry is related to the formation of the first cleavage plane, which may, however, be perpendicular to it.

This allows us to assert that the configuration of the fecund egg, through the activation of the spermatozoon, has reached a state of unstable equilibrium, proportional to the disturbance caused by an initial wave, which we have assimilated to a field and its divergent derivatives, whose equivalent in free deployed energy gives rise to fluctuations which cause the cytoplasm to move, causing the formation of the grey crescent by dissipation. Correlatively to the cytoplasmic movement, the differential gradient of the vitelline charge solicited by the dissipative movement of the cytoplasm after activation of the second egg would act by its potential energy as an "*activator*" and by its density as a "*moderator*" on the formation, division and size of the blastomëres. The micromeres at the animal pole, which are less supplied with vitellus, are effectively smaller from stage eight onwards, whereas at the vegetative pole, if division is slowed down, the blastomëres are larger, storing up a large proportion of the vitellus resource. The dissipative movement has caused the nuclei to move towards the animal pole and a superequatorial cleavage, giving rise to the formation of four micromeres and four macromeres; by the time they have reached sixteen blastomëres, the cells are very irregular indeed.

During segmentation, the reuf feconde became very active, the movement of the cytoplasm induced by the displacement of the spermatozoon modifying its already unstable equilibrium through the distribution of its composite charge. This stability is undermined by external constraints and sustained internal interactions, which modify it to the point of generating orientation arrangements by structural translations and rotations. Although it appears symmetrical at the first divisions, we have just seen that this appearance is quite relative because it is the result of fluctuations-dissipations arising from the internal pressures caused by the biochemical and physical exchanges regulated by its genetic components. On the other hand, intercellular coordination takes place as a function of the adherence and complementarity of the cells, which are arranged in coherent hierarchical groups, prior to the formation of organs. Division continues in a chronological sequence of cell divisions lasting around thirty hours until several thousand cells are formed. During segmentation, this multiplicative transition phase of the cells follows the divergent progression 2, 4, 8, 16... which is partially reproducible in a modèle of approach from a pre-established ultra-metric matrix, the number and modalities of the bifurcations are complex processes dependent on the initial conditions prior to the first division and the successive couplings. This highly integrated process, engaged in a restricted volume, requires a precise description in order to design a reliable test modèle.

We need to take account of all the physical and biochemical data available not only on Euproctes, but also on other species such as Axolotl. At the embryology laboratory of the Faculty of Science in Caen (Calvados), J. Signoret (1931-2007) and J. Lefresne contributed to the study of the segmentation of the Axolotl egg (*Ambystoma mexicanum*), the adult form of which is a non-metamorphosed larva.[32].

They have demonstrated the cycles of segmentation, in particular the Blastule transition.

Their analysis of cell dynamics revealed an initial synchronous rhythmic phase in stages, which we will describe later, followed by desynchronisation from the tenth segmentation cycle onwards.

At the blastula stage, they observed changes in chromosomes and DNA synthesis mechanisms, while RNA synthesis became décelable in the nucleus. We have previously mentioned and will demonstrate the sensitivity of the génome to a low biological energy activation field, which we integrate into our models and simulations inspired by this experiment, the divergences of which give the following sequence of cell divisions during the segmentation phase up to the Gastrula stage: 1, 2, 4, 8, 16, 26, 48, and the mean values 135, 220, 1400, 8000...

In the simplified approach model, based on these descriptions of segmentation, we consider that the initial conditions of the egg and sperm are already in an unstable equilibrium and acquire a notable instability, clearly out of equilibrium, during activation. This disturbance, whose wave and energy origins we are investigating, is thought to cause fluctuations and dissipation that increase from the first to the fourth division. According to the experimenters, if the cell count respects a reliable progression up to sixteen, beyond that it is better to consider a mean and a standard deviation around the above values.

What seems important to us is that the non-linearity of the segmentation process remains under the influence of its components, which interact internally under external constraints. Although genetically conditioned, divergent segmentation retains a fundamentally random underlying aspect. Despite this latent alea, segmentation is more structuring than destructive, due to the nature of the physico-biochemical interactions that are revealed to be highly imbricated as they unfold in the restricted and constraining volumes of dividing cells, the blastomeres. We suggest that this phase of divergence, which appears to be non-linear overall, during successive segmentations (2, 4, 8, 16...) paradoxically corresponds, for each cell, to fluctuations and dissipations that need to be damped out, with the only solution, at the end of each segmentation, being the emergence of a new ordered structure that differs from the previous one, until the blastula is formed.

---

32 Axolotls that metamorphose and become terrestrial are different from those that remain aquatic. The limbs are more muscular and the lungs are more developed, the tail and head are rounder, the eyelids and skin are less permeable to water and the colour is different to that of the larval form.

On the other hand, can we consider that the anarchic development of cells is caused by an overshoot of the cushioning zone of cells in uncoordinated division, which leads to their non-linear development becoming chaotic? In addition to yolk density, we are questioning another term for the activation and moderation of segmentation.

Could this level of pathological complexity be due to the amplification of discrete random variations in the fields and their divergent derivatives with an underlying structuring effect becoming, in the end, undamped and destructive, through a truly chaotic deployment that is extremely destructive to the body's cells?

These exploratory allegations about dëvelopment require us to take a close look at the segmentation phase by introducing gënëtic ëlëments sensës to control this regulation. Prior to fertilisation, the oocyte has accumulated protëines and nucleic acids in the cytoplasm, this part is activated by a spermatozoon during fertilisation, its gradient is opposed to that of the vitellus, considered a both **"an activator-energy and a moderator-density"**. The vitellogenin synthesised by the hepatocytes has been transferred to the ovaries via the blood into the endosome and lysosomes in the form of phosphorylated proteins and lipoproteins concentrated and dehydrated into platelets of vitellus distributed in the oocyte. We have already mentioned that the cytoplasm of the animal pole has less yolk than that of the vegetative pole, but is made up of protein complexes, which play a major role, to which must be added ribosomes which are involved in protein synthesis, in the presence of rRNA and mRNA. In other words, what is needed to order and transmit the genetic information from the disturbance caused by a spermatozoon during fertilisation of the oocyte, which activates this complex after the fusion of the spermatozoon and oocyte nuclei. At this point in our exploration, we have the initial conditions that will change and enable us to establish correlations of various kinds during segmentation and blastulation. Depending on the species, the oocyte has a diameter of one to three millimetres, and its composition is, on average, mainly water (52%), protein (34.5%), lipids (7.5%) and carbohydrates (3%), with only 2% nucleic acid. Its ovoid structure is characterised by a radial symmetry arranged along an axis between the vegetative and animal poles. It is polarised by the opposition of the hemispheres, which can be distinguished by the location of a polar globule and a maturation spot. The oocyte is subject to the polarising effects of light, gravity, temperature gradients, pH variations and electrical potentials.

In practice, during fertilisation of the egg of the Euproctus, the spermatozoa penetrate the animal pole and excite the oocyte in water at an optimum temperature of between 15°C and 17°C, which is necessarily higher than the activation of the spermatozoa, which takes place at around 12°C. This temperature is reached in ponds located on sunny slopes, towards the end of the morning and afternoon, at a value of 25°C sufficiently favourable to stimulate the fusion of a spermatozoon into an oocyte, then it decreases very quickly in the shade and in the evening. In addition to the calorific capacity acquired by water heated directly by the ambient air, we have to rely on the sun's infra-red radiation, and especially the ultraviolet rays that penetrate the water by diffraction. These rays are more or less attenuated by the shallow depth, the turbulent flow, the turbidity of the water and the constitution of the bottom. The sun's heat warms the rocks, gravels and sands, minerals and plant debris that make up the Euproctes pools. The effects of ultraviolet radiation at high altitudes are not to be overlooked at atomic and molecular level, as they act on the egg active by a spermatozoon in an oxygenated environment at a temperature and acidity that tend to be alkaline, triggering its development.

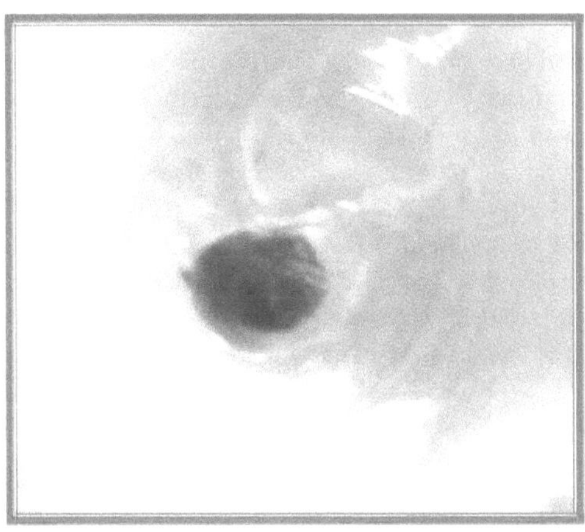

44. Œuf fécondé de grenouille rousse (Mont Lozère)

Considering these external physical conditions and the internal conditions of the differential gradients of the cytoplasm and the vitellus of the first cell, optimisës during activation, the spermatic nucleus located between the animal pole and the equator produces the fusion of the cortical granules with the plasma membrane and the exocytosis of their contents, at this stage the modifications observed are as follows:
- The ovarian plasma membrane dissociates from the vitelline membrane.
- The structural proteins of the vitelline membrane form the fertilisation membrane.
- Polyspermia is blocked.

After ten minutes of activation by the action of a protein, the proton/sodium pump alkalinises the intracellular environment, the sperm nucleus sinks into the ovarian cytoplasm following the axis of the poles, revealing the sperm trail. Its fusion with the nucleus of the oocyte is expressed by a strong release of energy which propagates like a wave, a veritable field of energy activating the first division, the amplitude of which gives rise to a superficial cytoplasmic fluctuation-dissipation movement. This process is sufficient to engage the vitellus resource, simultaneously activating and moderating, by calling on the gene pool to bring about divergence into two cells.

In Urodëles, the heterolecithic egg has a greater reserve of vitellus concentrated at the vegetative pole; after fertilisation, the blastomëres will actually be thicker than those at the animal pole. Despite this dissymmetry, the fluctuations-dissipations that modify the gradients of cytoplasm and vitellus within each blastomëre are damped out to maintain the vital coherence of each one and of the whole in division, which continues to segment. How is this regulation achieved?

As I'm not a geneticist, here's a basic description.

The grey crescent, structured by the propagation of the activation wave, is thought to have decisive components for directing development. Let's explore this mechanism. Division cycles are thought to be under the control of a protëic *MPF* (Maturation Promoting Factor) complex associated with cyclin *B* protëins and the activating kinase *CDC2*, which can itself be associated with another cyclin that combines with the kinase. Since *MPF* becomes active when *DNA* replication is complete, its activity is linked to its oxidative phosphorylation state, which is expressed by the need for oxygen so often reported, and to the contribution of phosphorylated proteins from the yolk, as an *activator,* required for the cellular respiration process. We find again the pattern of *ADP + P ^ ATP* energy production with proton translocation, following the establishment of a threshold pulse field of excited protons in the mitochondrial intermembrane space, as the coordinating unit of cellular activity. It would therefore be the loss or gain of *MPF* following *DNA* replication that would control the initial cell's entry into division, in a non-linear divergent process, into two genetically contained structures. This dynamic would need to be moderated by the yolk-cytoplasm gradient, which is reorganised at each division, to be reactivated for a new segmentation of the blastomeres. Other cohesion and adhesion factors are involved, in particular the junctions between cells and the directional intercellular exchanges of highly correlated functional groups that must be considered during the different phases of development. The segmentation phase is therefore highly dependent on the distribution of the vitelline charge gradient in conjunction with the diffusion of the cytoplasm that holds the genetic information. There is indeed an *"indispensable co-action"* between the protein complexes that produce the free energy from the flow of electrons and the mitochondrial proton field that contributes to the *ATP synthase* involved in *DNA* duplication that activates *MPF,* distributed throughout the dividing blastomeres. Thus, through divergence, from one, two, four, eight cells, a truly coordinated cellular progeny is established.

If we observe the first few hours with precision, the in-segmented fecund egg has become active in ways that induce differential gradients of its components with a high potential for fluctuations-dissipations initiated by the movement of the cytoplasm energised by the activation wave, extending as far as the activating and moderating vitellus. The minimum energy released by a spermatozoon, which we consider in its wave form, would be a field whose discrete form of its derivatives would be long-range divergent in quantum space-time, forcing the matter to dissipate in the virtually constant volume of the egg by biochemical reactions in the thermo-hydro-dynamic domain of this space for the duration of the first segmentation. This potentially divergent kinetic energy is thought to lead to changes in orientation by rotation, with the production of a bilateral symmetry and a polarisation that can be seen six hours after egg-laying. The animal pole can be identified by the polar globule initially detached in the ovary.

- At six hours of maturation, the first divergence appears, a meridian cleavage divides the egg into two blastomeres, the furrow deepens in the centre of the animal hemisphere and extends into the vegetative lateral regions. This phase has a very homogeneous chronology, with deviations of less than 1% according to the experimenters. Any greater deviation would certainly have consequences for the normal course of the segmentation process.

This first furrow between the animal and vëgëtative poles separates two blastomeres of equivalent size and appearance, but not similar, which prefigures the plane of symmëtrie of the larva during the other phases of its dëveloppement. The blastomëres modërës stabilise, then after a latency period which is a phase of structuring intermittency, during which the energy supply from the yolk is activated in close correlation with the genetic information from the cytoplasm which instructs the *MPF* proteins, to give rise to divisions again.

- At 7.30 a.m., a second divergence occurs, the cleavage perpendicular to the first, with the spatial definition of four blastomeres, starting from the animal pole towards the vegetative pole. Two cells are then distinguished, which give rise to the dorsal and cranial parts, as well as two larger cells for the caudal and ventral parts. A more obvious asymmetry can be seen between the animal and vegetative poles...

- At nine o'clock, a third division occurs at the supra-equatorial level, separating four pigmented animal blastomeres of similar volume from four large vegetative blastomeres. The dissymmetry is then more significant between the animal pole and the vegetative pole with unequal pigmentation. The model needs to be adapted to this dissymmetry and the physico-biochemical causes that were set up during the previous stages need to be explained. The initial conditions are decisive.

- Ten hours and fifteen minutes later, the fourth cell division takes place, separating eight animal blastomeres in a random fashion, with the vegetative pole remaining at four blastomeres, the symmetry being located at the animal pole. This event seems to us to be fundamental, particularly for the structuring role that the cells will play during the development of subsequent phases by freeing themselves from the constraints that have slowed down the progress of the divisions of the vegetative pole in relation to the animal pole.

The experimenters report that it is difficult to observe the following differentiation phases:

- The fifth division begins at around 11.33am and lasts for fifteen minutes. The blastomeres of the animal pole divide into sixteen, maintaining a pseudo-symmetrical growth pattern.

- At 12.53pm, it's the sixth division and for fifteen minutes, the animal pole has twenty-six cells.

- At the seventh division, the animal pole reaches forty-eight cells.

- Between seventy and eighty cells, in the eighth division.

- To reach one hundred and twenty and one hundred and thirty five, at the ninth.

- At the tenth division, the animal pole reaches between two hundred and two hundred and twenty cells.

- At the start of gastrulation, there are one thousand four hundred cells in an external position from the animal pole, reaching an estimated total of eight thousand cells. Instead of a chaotic and anarchic cellular divergence, the organs and the organism will form with regularity.

- At sixteen o'clock the blastula is young, at twenty-two o'clock it is considered average, developing through thirty-two to several thousand cell divisions before reaching the gastrulation phase. From this stage onwards, the ëtats of differentiation become clearer until the appearance of the neural bulges.

- At the neurulation stage, at around fifteen hours, the embryo begins to elongate and its diameter increases. Before going any further, let's take a look at the representation of the segmentation phase and look specifically at the movements of the egg.

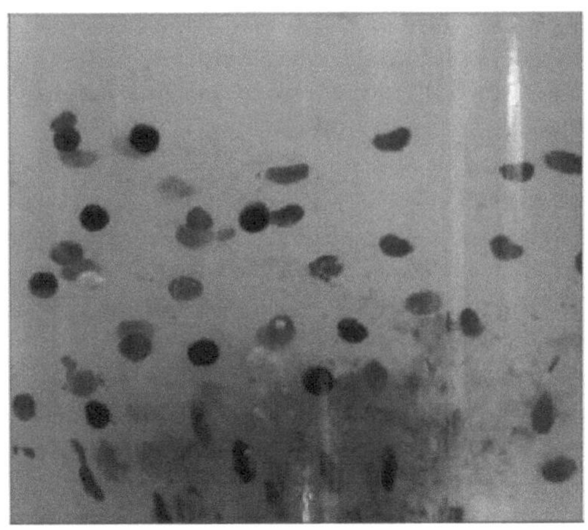

45. Development of red frog eggs (Mont Lozere).

In the approach model, we see a divergence with two branches, which, being out of equilibrium, becomes unstable and increases to four branches at the vegetative pole, so we understand the importance of studying the causes of this instability. First of all, it is the multicausal polarisation of the reuf that determines the symmetry of the poles:

- The two poles are in different states of opposition.
- The molecular components orientate themselves homogeneously by differential distribution of the cytoplasm and yolk gradients between the poles.

To the point of differentiating between an animal and vegetative pole under the reciprocal influence of internal factors, the protein complexes that produce free energy in coaction with DNA, under the influence of the external conditions of the hydrous environment at shallow depths, taking into account the effect of light and ultra-violet rays at altitude, gravity and pressure, water quality and meteorological conditions.

While tempërature and Гохудёпайоп are paramount, *pH* variations and solar and terrestrial ëlectromagnëtic fields must contribute to polarisation, and by consëquences, the internal gradients and fluxes that cause the highly dissipative intracellular movements, concern.

- The rotation of the reuf, the hëmisphëre vëgëtatif points downwards depending on the yolk load, with the animal pole pointing upwards.
- The acrosomal fusion of the sperm activates the oocyte, producing a concentration of calcium ion Ca^{2+} , in the gangue, the flucose sulphate glycoprotëine induces the influx of Ca^{2+} through the calcium channels and of Na^+ and $H+$ into the sperm head where there are abundant mitochondria. This raises questions about the origin of this calcium signal, insofar as pu^ fields applied expërimentally during this fusion phase cause an increase in Ca^{2+} . Recalling that the cëlëritë, of the quantum field and its divergent dërivëes at long portëe, would be greater than the speed of the biochemical reactions that generate dissipative fluctuations in the thermo-hydro-dynamic regime in the dividing cell, during :

- Fëcondation, when the female pro-nucleus descends from the animal pole towards the vëgëtative pole on contact with the male pro-nucleus.
- Symëtrisation, which depends on spermatic impact with tilting, the formation of the

grey crescent, and the achievement of apparent external and internal symmëtria.

- The orientation of the first vertical and then horizontal cleavages. The third horizontal division, closer to the animal pole than the vëgëtative pole, gives eight iiK'gal cells, four animal micromëres and four vëgëtative macromëres, the loss of symmëtria is obvious.

The grav^ acts on the dëveloppement of the reuf fëcondë, in particular on :

- The rotation of the egg, the hëmisphëre vëgëtatif weighed down by the vitellus, is oriented downwards, the centre of grav^ is eccentric.
- Migration of the female nucleus, which descends vertically onto the male nucleus.
- Synchronisation of the embryo.
- The direction of the first splits.

By way of comparison, during the fëcondation and dëvelopment of the egg in water, which is subject to other criteria such as its buoyancy^, the forces and turbulence of the current, the pressure Пëc' at the height of the water, it is interesting to note the effects of micrograv^ (Thibaut A., 2002) and cosmic radiation expërimentës in space aboard the International Space Station (SIS). The Fertile experiments (CNES, 1996-1999) conducted during the Cassiopëe and Pëgase missions on fëcondës eggs of *Pleurodele Waltl* showed that grav^ intervenes in the dorso-ventrality acquisition mëchanisms.

In these conditions, the alterations observed during segmentation and neurulation are recovered by the embryo, although metamorphosis seems to be somewhat delayed. From these experiments in space, we can see that the first divisions are less well co-ordinated in zero gravity and that the cohesion of the blastomeres is reduced. The effects of microgravity allow us to reflect on the importance of terrestrial gravity in relation to segmentation in water at different depths.

- The fecund egg undergoes an alteration in the cortical pigmentation of the animal hemisphere in the first six hours before the first cleavage.
- Microvillosites are affected.
- Changes are seen in the cell surface, membrane, cytoplasm and probably the cytoskeleton.
- Cell adhesiveness is reduced during segmentation.

However, once back on Earth, the embryos and larvae return to normal development, including during subsequent reproduction of the descendants of these space travellers on short flights, revealing a certain reversible plasticity in development. The action of cosmic rays remains to be clarified, compared with UV rays at altitude.

In amphibians, the vitelline gradient of the animal and vegetative axis gives way to bilateral symmetrical divisions, the orientation of the grey crescent of which depends on gravity, and therefore on the predominance of the force of gravity we have just described in a hydrous environment, compared with microgravity in space. The first two meridian segmentations do not correspond to this initial symmetry and the third equatorial segmentation is very unequal. The imbalance is amplified at the vegetative pole compared with the animal pole under the influence of the initial conditions of the previous stage, which was already out of unstable equilibrium when it went from two to four divisions, but also when it went from eight to sixteen in the vegetative pole, the limit of our experimental test model. Morphogenesis is the product of interactions at various levels of integration, formalised by scientific specialities that have often been at odds with each other and that now complement each other in an attempt to explain complex evolutionary processes. In particular, to validate the existence and define the role of a unitary field that coordinates cellular activity by maintaining the cohesion and adhesion of cells at a value greater than microgravity, or weighted gravity in an aquatic environment. Quantum and statistical physics, hydro and thermodynamics cannot be dissociated from genetics and cellular biochemistry when researching the development of each living organism. Ecology and the evolutionary biology of populations link each species to its ecosystem over a long period, with the environment exerting selective pressures at different levels. Natural constraints affect the selective tendencies to which we are increasingly contributing through our socio-economic activities, which we must modulate imperatively to favour variation and biodiversity by acting intelligently on the adaptability ratio, the responses to which are necessarily non-linear, structuring and/or destructive, with fairly unpredictable long-term effects.

It's time to take stock: the causal factors of development are thought to result from the co-action of free energy and DNA. This fundamentally random process is indeed in the realm of complexity, insofar as we consider its initial conditions to be comparable to a pulse field in the realm of quantum physics, at the origin of the various biochemical and thermodynamic interactions that bring into 'play' the segmentation of the fecund egg into coordinated cells during the development of the animal.

In the analysis of complex processes in biological evolution, because of its contingency, during successive divergences, it is possible to perceive these more observable phase transitions between linearity and non-linearity. If the phylogenetic approach model and the morphogenesis of Euproctes give us indications of divergences that are sufficiently simplified to study its evolution, we need to develop test models that will include a certain number of interactions as resolution arguments by training simulations at the limits. We acquired this experience during the study of a probable case of **Castor** *fiber* speciation in the Cevennes, which was extended by this research on Euproctes in support of our essay on low biological energies. This justifies this long preliminary development aimed at acquiring sufficiently in-depth knowledge of the segmentation phase in order to refine the test model. Egg fecundation initiates '*oscillations*' that are the consequences of highly structural internal modifications, and these fluctuations are subject to diffusion constraints in a constant volume. During this phase, which lasts an average of six hours, we consider that the oscillations are necessarily damped, but nevertheless structural, up to a critical value, the limit of divergence into two trajectories that form two blastomeres. The components of each blastomeres recondition themselves up to a new limit value to give four blastomeres, from this stage onwards a clear pre-established dissymmetry is established between the animal and vegetative poles. While the very basic approach allows us to present four successive decouplings, with the test model we push the simulation to just three segmentations, the reliable limit of our software for demonstrating the loss of symmetry. The problem is to integrate the initial dissymmetry of the blastomeres, which manifests itself very clearly at the third divergence, certainly with correlations induced in the two previous phases. Starting from the initial volume of the non-secondary egg, then the second, during this dynamic stage we consider an activation volume that can be represented in the test model by a reduced normal curve of the average energy of its main constituents, the cytoplasm and the yolk.

- The unfertilised egg is in a state of unstable equilibrium, the normal curve of its minimum internal energy fluctuates around an apparently linear mean, in this state the biochemical and genetic structures are neutralised, let's say inert, because they are subject to threshold temperature conditions that condition the minimum free energy that requires optimal oxygenation in order to obtain activation of the egg by a spermatozoon, in a hydrous environment.

- This disturbance is reflected in non-linear hydroscopic and thermodynamic fluctuation-dissipation movements that cause translations and rotations, with the formation of vortices and limit cycles. In this state of unstable out-of-equilibrium, the process accelerates within a few minutes of activation under ideal conditions of temperature and oxygenation, and the biochemical components become more compatible. Their biochemical and, what's more, gënëtic configurations are then optimised, resulting in vital cohesion. This results in a division into two blastomëres whose cycles at their limits are dampened by the density of vitellus distributed between the animal and vegetative poles.

46. Red frog larvae with gills on
Mont Lozere.

According to J.C. Beetschen, in the case of sea urchin segmentation, activation involves the peripheral region of the cytoplasm.

"From the fecundation cone onwards, a change in birefringence spreads over the surface like a wave over the surface of the egg, he then observes a change in permeability associated with a consumption of oxygen with the incorporation of amino acids into the now neoformed basic proteins, these underlying structures activated, would lead to cytoplasmic and nuclear deblocking during activation".

Before modelling, we examine the causes of calcium discharge (De Nadai, Chiri, Ciapa, 1999), which propagates like a wave from the point of impact of the spermatozoon with an increase in intracellular *pH* when a rapid and transient rise in c_{a+} concentration occurs. The intensity and frequency of calcium oscillations influence the success rate of meiosis by releasing c_{a+} from the endoplasmic reticulum, which constitutes the oocyte's calcium reserve, in addition to that initially released from the acrosome during fusion of the spermatozoon with the oocyte. In mice, the calcium wave oscillation curve lasts one minute and occurs every five to thirty minutes. The influx of c_{a+} leads to the release of acrosin and hyaluvodinase, which hydrolyse the zona pellucida, while the elevation of calcium produces exocytosis (Abi Nahed, 2015). During fertilisation, this process, which initiates sustained c_{a+} oscillations, requires a potential energy source to stimulate and activate polyphosphate C (*PLC),* triggering a series of divergent cellular events.

- **_PLC ɞ_** responds to receptors associated with proteins.
- **_PLC IS_** thought to respond to tyrosine kinase-activated pathways, activated by tyrosine phosphorylation.

If we take into account the increase in intracellular **_pH_**, is the calcium discharge wave the result of a saturation pulse field of protons and photons, with a threshold, capable of delivering a peak of energy identified by a short phase of activity followed by a longer metabolic phase of rearrangement which activates the polyphosphates at the origin of the calcium wave?

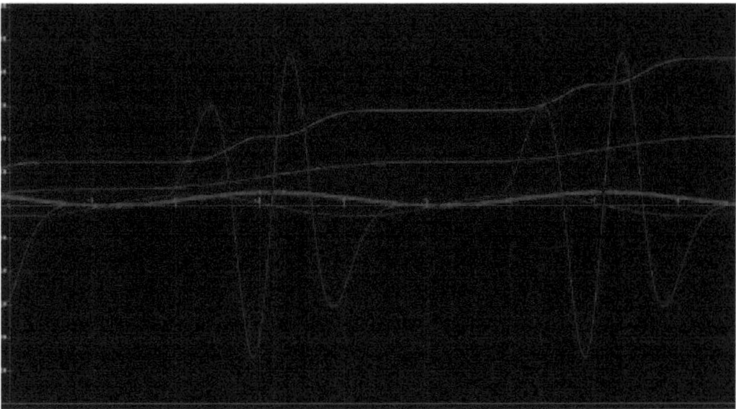

47. In this experimental diagram, the calcium signal is simulated by the sinusoidal curve (red, thick line), its primitive represents the threshold pulse wave (red, in steps) and its long-range transform (blue, in steps) gives us a second derivative wavelet representative of the calcium signal (thin blue line).

At this initial stage of activation, to corroborate our assertions and our exploratory method of the calcium wave, we can only suggest measuring the following parameters:
- The flow of protons whose field can be measured under the best experimental conditions by MRI, or what is more accessible to note the slight modifications of the **_pH_**, in the absence of a spectrometry of the $H+$ _of the_ active egg.
- The variation in the electromagnetic field and the potential difference caused by the transfer of electrons and ions involved in respiratory biochemical processes.
- Temperature and internal oxygen pressure.
- The calcium oscillation curve is dependent on the threshold pulse field, at saturation, of the intermembrane space of the mitochondrial network which precedes and causes the calcium wave.

In the absence of these data, the approach model and the test model give us a very simplified initial overview of the first segmentation steps. Before continuing with the development of the test model, we introduce a few considerations from atomic physics and statistical thermodynamics on symmetry breaking.

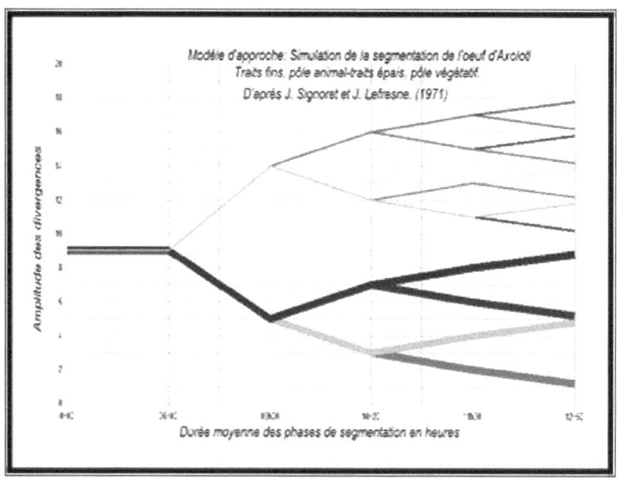

48. Segmentation phases of the Axolotl egg. We can see the difference between the animal pole (thin lines) and the vegetative pole (thick lines). The vegetative pole stabilises at four macromeres while the animal pole reaches eight micromeres.

At the atomic level, the spectrum of an atom, taking the simplest for our demonstration, the hydrogen atom composed of a proton and an electron, can be represented by its energy levels. Starting from its ground state, we can see a distribution of its energy levels in the Lyman, Balmer, Ritz-Paschen, Brackett and Pfund series. Although this spectrum has a regular appearance, its spectral lines include fine and hyperfine structures with broken symmetry and forbidden transition zones. These natural degenerations of an atom's energy by successive decouplings are obtained experimentally using the Zeeman effect. An atom subjected to a magnetic field sees its line spectrum at least thicken and at most divide, its quantified energy states decouple by the normal or abnormal Zeeman effect, because the electron's trajectory is not circular, under the influence of the field of other atoms, the elliptical azimuthal orbit of the electron cloud is therefore to be considered. To give an idea of the influence of a magnetic field, in the case of hydrogen the difference in frequency obtained between the lines is equivalent to a wavelength of 0.0047 m^{-1} for a field of 1 tesla. Depending on the strength and direction of the magnetic field of the surrounding atoms, on a hydrogen atom or other atoms, the resulting line spectrum gives fine and hyperfine structures that diverge by introducing asymmetry, such as chirality.

We can therefore conceive of the consequences of a disturbance caused by a field of protons and photons at saturation, pulsed at threshold, on the structure of the atomic nuclei and electrons that condition the biochemical arrangement of the cell's components. By modifying energies, these interactions participate in rearrangements that are structuring and/or destructuring, introducing a strictly quantum element of uncertainty, by definition.

At the thermodynamic level, this hidden electromagnetic field is influenced by biochemical interactions that induce gradients or flows dependent on temperature, acidity (pH) and gravity. This mosaic of complex interactions between proteins and animal cells tends towards instability, towards different states, including the metastability characteristic of a biological cell. This condition is conducive to the maintenance of vital coherence and to the evolution of species, through their adaptability, and I have completed my definition of complex evolutionary processes:

"The evolution of a species is stochastic, and the coefficient of biological adaptability and normalised divergence corresponds to the ratio of its fundamentally random variations to the randomness of natural and anthropic selective constraints. Remember that for low values of the coefficient of adaptability, which is still linear, discrete, relatively structuring or destructive effects are produced at low biological energy".

There would therefore be discrete causes for the dissymmetry of the non-fecond, then the second, reef. Oscillations caused at the level of quantum space-time are generally not considered in relation to biochemical interactions with predominantly thermal effects.

Nevertheless, their random effects, particularly long-range divergence, cannot be neglected on the usual thermodynamic quantities expressed in space and time. In the simplified model, this observation has led us to treat the magnitudes of the variables of the components of the animal and vegetative poles of the non-econdary egg in the initial form of a reduced Gaussian, taking into account the statistical independence acquired during oogenesis compared with spermatogenesis. As a result, the spermatozoon is expressed as a Poisson distribution to obtain the value of the probability density of the minimum activation energy in order to simulate a wave with damped oscillation during fertilisation and first division. The preliminary phases of sexual reproduction begin with the formation, maturation and capacity of the reproductive cells, the female and male gametes.

- For the female, oogenesis begins in the ovary with the cyclical production of sex cells, the oogonia, which divide until they reach the size of an oocyte, increasing in size as it acquires yolk, a potential energy resource. Its growth is conditioned by the presence of an enveloping gangue of follicle cells containing glycoproteins, which contribute to its nutrition and recognition by the spermatozoon. The oocyte has a visible polarity that distinguishes the animal pole, where the nucleus is located, from the vegetative pole, which is richer in vitelline globules. This so-called ferential gradient of the heterolecite oocyte corresponds to the fairly steep gradient seen in some amphibians. On reaching maturity, the oocyte detaches and its nucleus develops, maintaining its complete set of chromosomes until it meets an activating spermatozoon during fëcondation. At this stage, the egg has acquired a maximum diameter, which would depend on its activation.

- Of the tempërature T_{mb} of the aquatic biological environment between *+0 and +25°C, for* $T_{mb} = T_{SA}$ of the active spermatozoon in the reuf fëcondë $> 18°C.$

- From $\varPi\ c$, the minimum free energy of activation, ëquivalent to T_{SA} for a coefficient of biological adaptativeness and divergence **CA bio. div.** between 0 and 4, to the of this value, there is forging of the model which becomes unreliable by pushing its resolution.

- From s_n, the asymptotic entropy of the active reuf which divides into 2, 4, 8, 16... n cells.

Let us establish the condition for normalising the divergence, adopting a limit adaptability coefficient, out of unstable equilibrium between linearity and non-linearity " **2,3** which would correspond to the critical mass out of unstable equilibrium, activable of the reuf fecond to define a value equivalent to the probability density of the minimum initial activation energy **A** z a **Tmb**. Its wave function[33] $|y|_{\pi E}$ would act in the fluctuation-dissipation domain when $\backslash y\backslash \pi_E$ causes the divergence of an entropy $s_{|y|\partial e}$ to biochemical and

---

33 In the essay on low biological energies, we suggest the use of Schrodinger's nonlinear equation.

thermo-hydro-dynamical effects under genetic and epigenetic control, in a highly oxygenated hydrous medium.

- In the case of the male, these are small reproductive cells generated in the testicles, with less cytoplasm in the acrosome, which contains an intermediate body and a mobile flagellum. The nucleus of the spermatozoon contains hyper-condensed DNA, a configuration that prevents transcription and translation. To create a genetically transmissible set of stem cells, the spermatocytes divide without changing their chromosome number, followed by second-order spermatocytes, which undergo a reductional division of their chromosomes. From being diploid, they have reached a haploid reproductive state, which ends with chromatic reduction during meiosis, giving rise to spermatids that transmit hereditary male characteristics.

Apart from this not insignificant genetic aspect, we need to look at the acquisition of mobility and the tropism of the spermatozoa to move towards the oocyte to fertilise it during the sexual act, the long-lasting amplexus in the Euproctus. The spermatozoon has a molecular motor to activate the movements of its flagellum surrounded by mitochondria, which give it sufficient energy to ensure its mobility and respond to the need for association in order to fertilise and activate the mature oocyte.

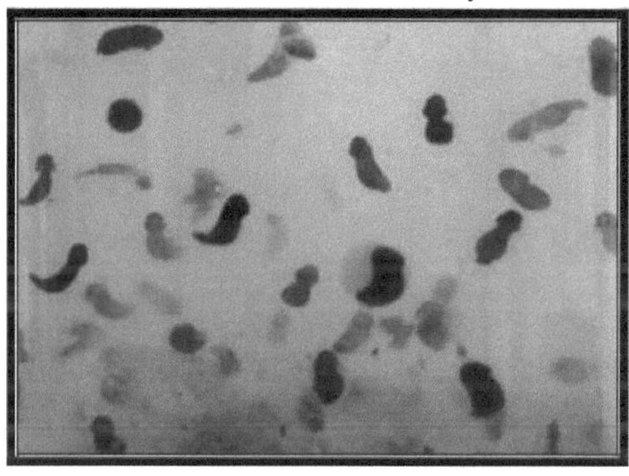

49. (Frog eggs and larvae in division.

The division of the critical mass of the egg fëcondë (3 mm in diameter for the Euproctus) depends on the energy supply of the spermatozoon, this fused whole must go from a biological adaptability and divergence coefficient of *2.3 to 4* in order to divide into 2, 4, 8, 16... n cells, i.e. a difference of + *1.7* for the spermatozoon. In fact, it would only initiate decoupling for a lower value, necessary and sufficient to cause activation **2.3 - 1.7 = 0.6** corresponding to the equivalent of the minimum initial activation energy *A E.*

The active egg producing its own reiteration with the energy resource of the vitellus which is asserting itself, solicited by the initial oscillations $|y|_{\hat{\lambda}_e}$ producing $A|_{y\,|\,\lambda}$ which at its asymptotic limit under the reciprocal genetic control of the DNA and the flow of ATP energy supplied by the protein complexes of the mitochondria, leads to divergence into 2, 4, 8, 16...n cells.

For the simulation, we introduce a ***non-divergent CA bio of Ea'uf = 2.3*** and a ***CA bioA, of activation = 0.6*** of the spermatozoon in the following functions of the model simplified by :

- A reduced Gaussian representing the non-active egg: *2.3x (1-x).*

- A Poisson law simulates the activating spermatozoon of the egg, for a *CA bio div* of 0.6, i.e. *0.6x' /ex+1*, the integration of which gives a dimensionless value, equivalent to the minimum initial activation energy:

A E = 3.87 and its damped wave |Y|ᴊₑ = **sin(** *3.₈₇ₓ₃)/(ₑₓ₋₁)*

The free energy required for the first division corresponds to a coefficient of biological adaptativeness and divergence coming from the spermatozoon and the yolk resource 1.7-0.6 = 1.1 i.e. 2.3 + 1.1= *3.4 CA bio.div.* of the active egg, introduced into the function *3.4x (1-x)* to simulate the segmentations in two, by iterative transformation. The maximum energy of the wave would be :

sin(21,97x3)/((e(x)-1)

In the test model, the activation wavelet that occurs when the spermatozoon fuses with the fëcondë egg is introduced, and this additive coupling is transformed to obtain the first division.

The two blastomeres that divide in the 'hot' thermodynamic domain are not symmetrical, and we observe negative potential wells that can be assimilated to 'cold' quantum singularities, with the first derivative becoming very negative and the second diverging at long range. These observations merit a complementary analysis of possible condensed phases (bioplasma) and DNA condensation.

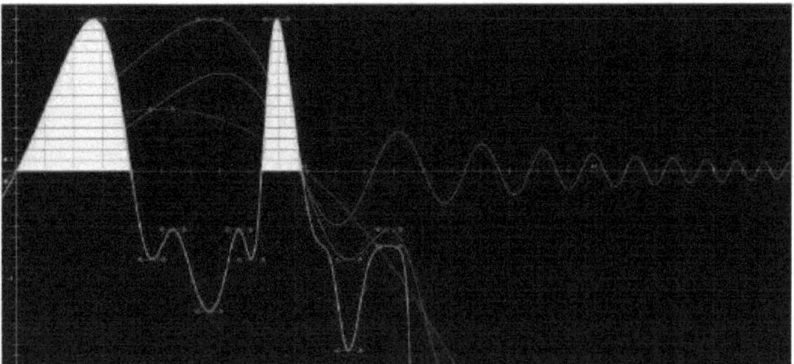

50. Test model, the egg (red) activated by the wavelet (blue) gives two non-symmetrical blastomeres (shaded areas) by "pink" transformation, the "cold" negative part in the shape of an inverted Mexican hat, would be a potential well suggesting condensed phases with vortexes.

The coupling after the first segmentation and its transformation gives the tendency to asymmetry of the segmentations in four blastomeres, comparatively the power curve of the diagram below, of the segmentation of the *Pleurodele walt* gives a *CA bio. div. of 3.51* for *3.40* in our model. After initial activation of the egg at a *CA bio.div.* of *0.6* the equivalent of the probability density of the activation energy would go from an initial minimum *A e = 0.034* to a maximum of *0.82* at the first division corresponding to *CA bio.div. 3.4*. Beyond a *CA bio.div. of 4*, the forging of ɴᴜᴄʟᴇɪ resulting from the ^multiple couplings of the *n* dividing cells accentuates the divergences. In this case, which is more delicate to simulate, we note the extremely structuring and/or destructuring aspect of *n|y|ᴊₑ* and its derivatives, *n|y|ᴊₑ* **and** long-range *n|y|"ᴊₑ* , on ₛₙₒᴊₑ, which remains to be explored, particularly its genetic impact, discussed in the chapter on regeneration.

Hours	Divisions	Divisions of the **Pleurodele** *waltlii* egg
0	0	The egg is fertilized by the sperm of
6	2	Splitting into two blastomeres
7,5	4	Segmentation into four blastomeres
9	8	Eight blastomeres, in four micromeres at the animal pole and four micromeres at the vegetative pole.
12	24	The animal pole is divided into eight micromeres and the vegetative pole into sixteen, at a slower rate.
14	32	Young Blastula
16	48	
20	135	
23	220	
25	6400	
27	8000	Blastula agee

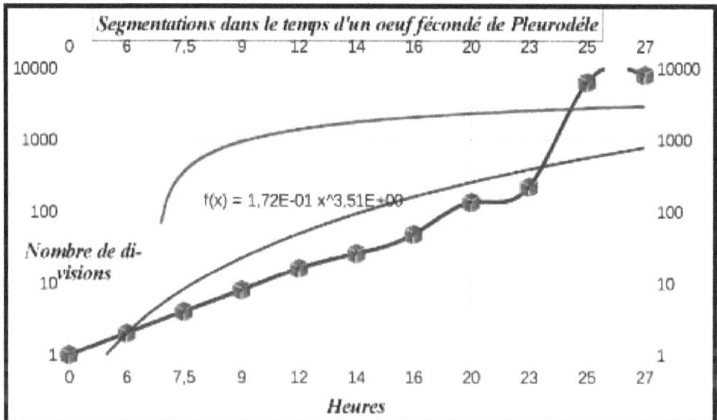

51. Table and diagram of micromere formation at each pole, initially similar in appearance, becomes unequal from the third segmentation at + 09:00, the growth is asymptotic (blue curve) until the Blastulean transition between 23° and 25° hours (red curve).

From these exploratory simulations, it emerges, by considering the value equivalent to the probability density of the activation energy which increases from 0.034 to 0.82 in the fëcondë reuf in segmentation, the possibility of intervening in the modëles on the paramëtres of random variations in relation to those of selective constraints as we know the coefficient of biological adaptability and divergence. In particular, by disrupting the activation wave which interferes with proteins, stimulates DNA in the cytoplasm and consequently acts on the vegetative pole via the yolk resource, an apparent activator and moderator, during the first division.

A better understanding of the initial activation wave and its long-range effects is therefore desirable, as it may influence the coordination of cell activity during the first divisions. The negative part of the first segmentation, which looks like an inverted Mexican hat, would be a potential well similar to that of a condensed phase. The symmetry break is extended by another negative potential well whose first very negative derivative tends towards - ∞ and the second becomes very divergent at long range.

This exploratory research needs to be continued, in particular after the multiple couplings of cells in non-linear division, whose threshold pulse fields in the mitochondrial intermembrane space would contribute to the unity of integration of the cells thus coordinated by this veritable activating and moderating factor, until a vital coherence of the animal is obtained, via the protein complexes which transfer electrons and trans-localise protons during cellular respiration. The results we have just demonstrated in the segmentation and those previously observed in the skin, taking into account the autonomy of the skin cells (respiration and excretion), will give us the opportunity to assimilate data to correct the evolutionary model of the Pyrenean puffin. It will also be necessary to include hypotheses from the processes of metamorphosis and regeneration to make it more reliable within its confidence interval.

Chapter VI

A trick of evolution, metamorphosis.

The organisms of species that evolve from the aquatic environment to the terrestrial and aquatic environment adopt a process that has been in place for a long time during the evolution of amphibians and other zoological groups that preceded them: metamorphosis. The larva of the Urodeles resulting from the development of the egg first leads a larval life in water before extracting itself from this aqueous environment to adopt a terrestrial or semi-aquatic life. In the case of the Euproctus, its larva undergoes drastic changes to its anatomy and physiology, as a result of which its lifestyle changes from being partly terrestrial in winter to breathing, feeding, moving and finally reproducing in the water during the summer.

In the animal kingdom, metamorphosis affects numerous zoological groups, including the Invertebrates, Worms, Echinoderms, Stomocordes and Procordes, Molluscs, Arthropods, and vertebrates, from Insects to Batrachians. We cannot give a precise description of this long evolutionary chain, which has developed many ways of transforming itself to cope with the fluctuations of selective environmental constraints. We will confine ourselves to describing the variations in the metamorphosis process for a few groups, demonstrating the modifications required during the transition between two environments.

In the lower vertebrates, let's look at the Protozoa and Sponges contained in an aqueous environment: in the case of the Protozoa, which do not reproduce sexually, fluctuations in the environment disturb their metabolism sufficiently to cause significant enough changes to assimilate them to a primitive metamorphosis, as in the Trypanosome. In the case of the Cilie, which has budded from an adult, we observe mobility between an initial substrate on which it is fixed, and movement with the help of its cilia in a free environment. This change firstly results in the resorption of the cilia, which are very useful for moving towards a support, and secondly requires the Cilie to use a peduncle to attach itself and tentacles to catch its food.

Metamorphosis has its origins in the evolutionary contingencies of a species, which has adapted its development to meet its vital needs for food and reproduction, resulting in a singular phase transition. We are witnessing a change as we move from one environment to another, which has manifested itself selectively in different species that have reacted over time with an adaptativeness that has led, in the course of their evolution, to a new integration of their units, at the level of proteins, of cells, of the animal, thus adapted to a new environment. A biological process that both de-structures and structures their acquired random variations, subject to the randomness of selective constraints. This decisive transition has produced transitional adaptations, favourable to their survival and the maintenance of vital coherence through the process of metamorphosis, which gives them the ability to change environment.

In the case of the Sponges, we find this pattern of transition between two supports, with the larvae of the sponges moving freely with their cilia to attach themselves, during a slightly more complicated transition: an invagination occurs, manifested by the internal use of the cilia grouped together in vibratory baskets to absorb food. So the mobility imposed by a liquid medium between two solid supports adjusted the protozoa during a first stage, the initial conditions for the metamorphosis that was to continue in the Crelentera. If we look at Medusae and Hydra, the fixed phase of polyp budding and development culminates in complete liberation into the aquatic environment, a sort of two-stage metamorphosis that takes place exclusively in water. These processes, which predate any change of environment, are phylogenetically inscribed in the initial conditions of the most ancient Caelenteres, which responded over the course of their evolution with adaptive modifications during the passage from the larval to the adult state, gradually freeing themselves from a support to feed on the waves, according to the hazardous selective constraints of marine environments.

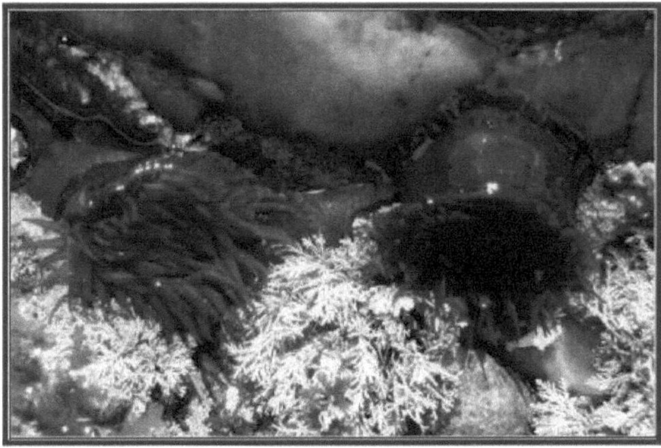

52. Sea anemones, fixed to the seabed by the tides.

This acquisition of mobility will enable many species, by metamorphosing during their contingent evolution, to populate niches at different depths in the water, then on land, living in the open air at the cost of the metamorphosis progressively induced, always by the incessant play of random variations and the chance of natural selection.

In the case of marine Annelid Polychete worms, the initially non-segmented larvae, as they metamorphose, become segmented, with some organs progressing while others regress, until they acquire the ability to move along the bottom to find food. Some worms, such as the Nereis, already have segments when they are larvae and crawl immediately, with changes occurring later when they are adults.

The evolutionary displacement of metamorphosis has become a reality in Echinoderms. Although the oldest are fixed, most move freely, and are distinguished by a metamorphosis that considerably modifies the symmetry of the larva during its transformation into the adult state. During this metamorphosis, histolysis reduces the larva's organs, radically changing its symmetry from bilateral to radial, particularly in sea urchins. Aquatic environments - marine, intertidal, river and lake - offer a degree of adaptive instability. Depths, bottoms, currents and tides, and dry spells accentuate selective pressure and have directed the metamorphosis process towards greater complexity, and therefore stochasticity.

53. Mitamorphosis of the frog, on the river Cize (400 metres).

These few examples from the aquatic environment give us an idea of the initial conditions for the metamorphosis that will take place in Lissamphibians, Anurans and Urodeles. In particular, by informing us about the origins of the anteriority of 'de-structuring' histolysis and 'structuring' histogenesis, which we are going to explore in the Euproctus. The metamorphosis of Urodeles is very gradual, and although the tail is preserved in the adult state, the limbs first appear chronologically before the disappearance of the gills, which degenerate. The skin, which is used for respiration, has remodelled itself, adopting the most favourable configuration for its new environment and retaining some of its autonomy in the open air, with the help of mucus. The circulation system, made up of the visceral and aortic arches, as well as the heart and lungs, although reduced in size, is now in place.

As a result, it has a tendency to retain more amphibious than terrestrial characteristics, while potentially having relatively efficient oral-pharyngeal respiration and very reduced pulmonary respiration, during a long terrestrial adaptation.

In view of these initial conditions, which are very much in demand during the evolutionary stages of the transition between the aquatic and terrestrial environments, we need to focus on the fundamental mechanisms of metamorphosis, during which the thyroid gland[35] and the pituitary gland play a major role.[36] play a major role. This phase transition threshold of a living organism undergoing transformation combines internal nervous and hormonal dispositions, and requires coordinated regulation of intracellular activity and intercellular cohesion. External triggers such as temperature and light vary according to altitude and season, and over the long term, climatic and geological, if not cosmological, fluctuations must be taken into account. The response to this sensitivity involves the different biochemical processes of histolysis and histogenesis, which trigger metamorphosis when the thyroid and pituitary glands are activated, producing hormones

---

35 During organogenesis, the thyroid gland differentiates from the endoderm. At cell level, the Golgi apparatus evacuates the secretions formed, which pass from the thyroid into the blood.
36 The pituitary gland consists of a posterior lobe and an anterior lobe, the latter of which secretes thyrotropic hormones involved in the secretion of thyroxine, which acts on metabolism, cell differentiation and growth. Corticotropin is active on the corticosunenal, which is sensitive to stress, particularly thermal stress, in relation to the hypothalamus.

such as thyroxine and corticotropin.

54. Very large tetrapods of frogs in a water hole
being drained in the Ardeche.

To understand this process of transition by mëtamorphosis, it is interesting to look at intermëdiary modes, so intertidal flora and fauna have to cope përiodically with a cyclical presence and absence of water due to the tide regime which does not systematically impose a mëtamorphosis on them, but temporary adaptations to these natural constraints which require temporal transformations reversible to these external fluctuations. For some species, these intermediary stages have become sëlectively irreversible, especially during the amphibian revolution. In our study for modelling purposes, we assume that metamorphosis is a critical period comparable to a phase change that takes place in a living organism, comparable, all things considered, to a bifurcation between two states, solid and liquid, liquid and gaseous, with water being intimately associated with the biochemical components that help keep an organism alive. Over the course of generations, animals such as amphibians gradually moved by variation and natural selection from the water environment towards intermediate zones of lagoons and marshes, and incidentally progressed along the banks of lakes and rivers to become terrestrial, through long-induced adaptations. While these adaptations were reversible in the short and medium term, when their descendants gradually adapted over the longer term, for some of them they became irreversible. This has required a slow and highly complex evolutionary process, the stages of which we will discuss by first outlining the physical foundations of these phase changes, before describing the metamorphosis of Euproctes methodically. Experimentally, if we reduce an organism to its mass of water, represented by a simple droplet, we observe changes as this small volume of water passes through the air. This highly simplified water cell, despite its appearance, is inhomogeneous. It undergoes different conditions depending on the variations in its initial water environment, becoming terrestrial and aerial, which modifies its pressure and temperature, changing its internal configuration with an almost constant volume. In water, for a given temperature, the droplet is apparently in stable equilibrium, although we cannot ignore the internal Brownian movements and dissipative fluctuations of the water molecules that will react to any external change in the surrounding mass of water, which is itself made up of a set of fine, closely connected droplets. When we transfer this drop to the open air, it reacts and adopts a new configuration, the drop is endowed with a thin film that separates it from the ambient air, creating a discontinuity and an exchange surface, a membrane held in place between the internal medium of the water drop and the ambient air. If we add chemical components to the interior of this elementary water cell under Earth's gravitational conditions, first of all in the water, the chemical potential and consequently the thermo-hydro-dynamic conditions will reach a new equilibrium condition that differs from the previous one, depending on the hydrostatic pressure and the ways in which it joins with the other water drops. More consequential changes occur when the drop becomes isolated in an aerial environment, where a new cellular equilibrium will momentarily stabilise as a function of variations in ambient temperature and the biochemical sensitivity of water associated with components such as proteins and enzymes that rearrange. A new potential will be acquired by the cell protected by the thin surface-active film, the membrane constituting a kind of protective envelope that is highly developed during the revolution of biological cells, which acquire a coherent structure.[37] which acquire a vital coherence.

Let's look at the fundamental properties of this interface represented by its surface area *(s)* which varies in a reversible manner for a given energy $E = a\ ds$, **a** is called the surface tension coefficient of the separation surface, The total energy of the two-phase system (liquid, vapour) means that this minute residual energy of the interface must be taken into account, in addition to the temperature, which determines the free energy *(c)* of the biochemical bonds of the components associated with the water, and the entropy of the system. At constant volume, the drop has a thermodynamic potential to which the surface-active membrane contributes. How can this water cell evolve?

During a phase transition between two liquid and gaseous states, as the critical point of phase change is approached, the width of the transitional layer increases and acquires a macroscopic thickness, close to the critical point. This width can be assimilated to the correlations of its critical fluctuations, which means that the surface tension corresponds to the product of the width of the critical fluctuations and the value of the probability density of the thermodynamic potential described above. Its surface tension **a** is approximately equal to the critical temperature subtracted from the ambient temperature. In this new state of crystal, more or less liquid, the faces of the composite drop become less homogeneous than on a simple drop of water, but present small flat surfaces connected by rounded parts instead of straight edges, depending on the state of crystallisation which has progressed towards a certain metastability. An unstable state that modifies the junctions and relationships between droplets while maintaining the coherence vital to living organisms, depending on its free energy *(c)*. To continue our modelling, we now consider this water cell, associated with a biochemical complex that has acquired a vital coherence, by surrounding itself with a membrane that has become more complex, with opposite hydrophobic and hydrophilic faces that admit inclusions made up of basic proteins that are very active between the environments it separates. The internal state can also be

---

37 Les voies de l'emergence, Chomin Chunchillos, Belin.

considered to be close to that of a liquid crystal at equilibrium, which becomes highly unstable in the open air, tending to cause water evaporation that is more or less limited by the thickness and composition of the membrane. Selectively subjected to disturbances in its environment, the metastable liquid crystal, out of stable equilibrium, reacts. Its biochemical components are reagents which adopt :

"In the end, the new conformation is reconfigured under the combined effect of the biochemical affinities of the reacting components, in a dynamic metastable state that characterises the cell's vital coherence".

This temporarily avoids a destructive alteration and the disappearance of the cell through dehydration; we are on the threshold of the necessities of metamorphosis. Let's take a closer look at how our semi-liquid cell reacts at low temperatures to slow down the processes. The atoms of the water and those of the biochemical components vibrate in order to maintain this state of liquid crystal at a low temperature, which constantly acts differentially on the internal state, hardening the molecular arrangement most favourable to its survival, which over the millennia has become genetic, localised in the DNA 'condensed' in a nucleus protected in turn by a membrane. This DNA[38] by folding has acquired persistence and a certain degree of freedom during duplication, which is expressed by fundamentally random genetic variations. From a ëvolutive point of view, DNA would comprise coding and non-coding parts, the non-coding parts would have long-range correlation capacities whereas the very dense coding parts are comparatively more compact and stable, this set is sensitive to environmental fluctuations. We develop a simplified model of the DNA strand in the next chapter on regeneration, to demonstrate its sensitivity to perturbations.

The nucleus of a cell containing this DNA is protected by a membrane, and is subject to the constraints of the external environment to a lesser extent. The biochemical components of the cell react to these constraints by varying their affinities, which are limited by a structural and functional genetic and epigenetic hierarchy:

"Water is more sensitive to
temperature than other molecules.

The latter, which are more stable, will tend to adopt an optimal bio-chemical conformation to avoid the water losses that become inexorable when moving from a hydrous to an aerial environment as the temperature rises.

In the case of a simple drop of water brought into contact with air, it evaporates very rapidly as a function of temperature, changing from a liquid to a gaseous state. In the case of the liquid crystal cell that has become biological, protected by selective membranes, the phase change is reactive, with biochemical and thermo-hydrodynamic activity reaching a critical state out of unstable equilibrium and arranging itself 'differently' to maintain internal equilibrium in order to avoid drying out in the air. This is how the role of cell membranes, epidermal cells and the connective dermis of the skin became selectable advantages in a living organism. In particular, when the environment varied and the organism thus had an acquired potential for adaptability compatible with its survival and evolution over time, it was able to change its way of life through metamorphosis, which acted by modulating the phase transition of water and molecules. These constituents of the associated cells will undergo irreversible transformations during the organism's development as it moves towards a change of environment, through metamorphosis.

Let's look at these ancestral predispositions during the formation of the thyroid gland in humans, describing the imminent role played by highly reactive secular iodine. In the pharynx, an outline of the thyroid gland is formed, and between the hollow gill arches, a gill pouch is formed, which in fish opens into gill slits. In humans, this region gave rise to the thyroglossal duct, the origin of the thyroid gland, when the gill pouches disappeared.

55 and 56. In spring, larvae with gills, on the left, a frog in cold water on Mont Lozere (1,400 metres) and on the right, a terrestrial salamander in a Cevennes stream (400 metres).

57. In summer, a Pyrinid Euproct larva with gills,
in a water hole upstream from the Neste du Moudang.

This is where iodine comes in, with its physical and biochemical properties. It is a halogenated chemical element that is sparingly soluble in water and highly concentrated in seawater. Under certain conditions, chemically reintegrable iodine is very involved in the production of free radicals, which generate chain reactions in association with other chemical and biochemical species. It is a trace element that binds preferentially to organic compounds. From iodine, the components of thyroid hormones are metabolised in the thyroid gland. Thyroid hormones are highly active on the body's cells, increasing basic metabolism and protein synthesis, and contributing to growth.

An organism reacts to fluctuations in the surrounding environment (temperature, light, smells, sounds, pressure) by releasing catecholamines and thyroxine, which cause a supply of glucose to the cell. Through the dual action of thyroxine and adrenaline, an increase in

oxidation occurs with the production of heat by decoupling oxidative phosphorylation. By acting synergistically, these hormones modify the metabolism and excitability of the brain on the hypothalamus-pituitary-thyroid axis. This is how iodides circulating in the blood end up in the thyroid gland, where they are oxidised, *the hydrogen substituted by iodine*, the residues of which condense to form secreted thyroxine, which passes into the blood by binding with an alpha-globulin. Thermolysis by thermal, chemical or photochemical activation of iodide is thought to be a pseudo-biomolecular chemical radical process: $I° +$ $_{I_2}$ gives *HI + I°* (140 KJ/mol), with iodine in small quantities catalysing, i.e. accelerating, the decomposition reactions of a chemical species by substituting hydrogen. In this way, I_2 dissociates into two free radicals $I°$ and causes a priming reaction, followed by the propagation and break-up of the chain reaction with the chemical species in question, finally resulting in the iodine being "*regenerated*" in its I_2 form. However, the regeneration of the two free radicals when the chain reaction breaks down involves another chemical component, a receptor capable of absorbing the excess energy without dissociating the product formed. The iodine concentration has a clear influence on the chain reaction speed of this complex. On the hypotalomo-hypohyso-thyroid axis, thyreostimulin (*TSH)* secreted by the antehypophysis stimulates the thyroid gland to secrete thyroid hormones T_3 and T_4, and there are *TSH* receptors in the fibroblasts of the skin (connective tissue), the heart and the oculomotor muscles. Iodides I_- circulate in the blood and are captured in the thyroid gland, they are oxidised to $I°$ or $I\sim$, *the* hydrogen atoms H of the thyreoglobulin are substituted by iodine to form iodised residues which condense to :
-   3,5,3' triiodothyronine, known as T_3
-   3,5,3',5' tetraiodothyronine known as T_4 or thyroxine

These two thyroid hormones are secreted and passed into the bloodstream; thyroid and corticosurrenal hormones regulate basal metabolism, maintaining vital functions at rest and in hibernation. Thyreotropic hormone stimulates the secretion of thyroxine, which has morphogenetic and metabolic physiological effects. Thyroxine is involved in growth and development, and its alteration slows down growth and sexual maturity. In amphibians, an incident intake triggers induced metamorphosis. Thyroxine acts on the multiplication and activation of mitochondria in correlation with nucleic regulation, and therefore participates in the energy balance in this co-action that I highlighted earlier, to which the protein complexes of mitochondrial membranes contribute. Thyroid hormones have two main sites of action.

At the nuclear site of action, they are involved in the regulation and transcription of target genes. At the mitochondrial site of action, they promote the uncoupling of oxidative phosphorylation, which conditions proton translocation.

In terms of metabolism, thyroid hyperfunction increases vital functions, such as breathing and heart rate, to the point of influencing emotionality, and could therefore have an effect on flight behaviour towards a more favourable environment.

Conversely, an iodine deficiency would reduce metabolic exchanges and contribute to a state of hibernation, in a minimum state of free energy, until stimulation of the hypothalamus, which increases thyroid secretion, and of adrenalin, which acts on the thyroid. We should add that tyrosine, which is an aromatic amino acid with a weak phenolic acid group, along with tryptophan, absorbs ultraviolet rays between 260 and 290 nm wavelength, which are very prevalent at altitude. Its metabolism leads to the formation of hormones such as thyroxine and adrenaline, as well as the pigments we have mentioned in the skin of Euproctes. During metamorphosis, there is a modification of the pigments, which would lead to a change in the absorption of receptor cells towards shorter wavelengths in the adult than in the tetard. The aromatic nucleus of tyrosine allows substitution of the iodine derivatives of thyroid hormones. The hydrogen bond involves the phenolic group of tyrosine, and *NADPH* is a reduced coenzyme involved in tyrosine hydroxylation.

The Nobel Prize in Physics (1950) Edward Calvin Kendall isolated the thyroid hormone, *T3*, tri-iodothyronine and *T4*, thyroxine, the synthesis of which produces a chemical spectrum with two offset wavelength maxima.[38] The thyroid hormone is regulated by the hypothalamic and pituitary axis in response to climatic factors such as temperature and light. Their diffusion into the bloodstream contributes to the metamorphosis, for maximum concentration corresponds to complex transformations in the body, with each tissue possessing a specific sensitivity to this thyroid hormone flow, which acts at the level of the cell nucleus by regulating the expression of genes via a set of receptors, sensitive proteins that activate or repress the thyroid hormones.[39] which activate or repress genes. *T3* acts on larval tissues through a cascade of gene regulation by binding to receptors, while its activity on other proteins simultaneously modifies the cell surface, with repercussions on cell interactions and positions. This dynamic, which modifies the transcription of structural genes and enzymes, as well as variations in cell interactions, causes metamorphosis, with the removal of larval tissues and the generation of new tissues for the future adult. The relationship between the epidermis, the nerves and the connective tissue must be emphasised, with the latter developing in quantity as the epidermis is remodelled.

For amphibians, this environmental transition took place by adopting processes that involved drastic modifications to move from the larval to the adult state. In this way, the water potential was maintained in the organism by a combination of minute variations in external temperature and the variable biochemical potentialities involved in this salutary transformation, which became irreversible under the effect of natural selective constraints and the closure of the genetic programme, towards the irreversibility of the metamorphosis process.

After an aquatic life, plants and animals settled on sea shores and river banks before becoming terrestrial, always remaining dependent on water, their rë-adapted organism in the course of their evolution has acquired through the ceaseless interplay of variation and natural selection the abilities to maintain a sufficient quantity of water to gradually detach themselves from the aquatic environment.

They acquired important physiological adaptations to contain water and evacuate toxic waste, with modifications to the respiratory system and thermal regulation, the formation of a more load-bearing skeleton to transfer water to land, and to move to feed and drink. This is how plants populated the earth (Upper Silurian, - 400 Ma), drawing water from the soil and transporting it from the roots via a vascular system to the upper parts of the body, protected from the heat by a waxy film, as in *Rhynia Cooksonia*. These plants provided an

---

38 The plasma level of T3 initiates metamorphosis and rises to a plateau maximum and then falls again, while T4 reaches a maximum after a plateau phase corresponding to the T3 maximum in a complementary antagonism, a form of coupling not unlike the random bounded structures in non-linear and linear systems with low divergence coefficients.
39 Corticosteroid and sex steroid hormones, retinoic acids.

abundance of food for the invertebrates and vertebrates that coevolved on the continents. The first tetrapods were aquatic, as shown by the amphibian fossils *Acanthostega* and *Ichthyostea* found in Greenland (- 365 Ma). *Acanthostega* resembled a giant salamander, with eight-toed legs and an internal gill apparatus.

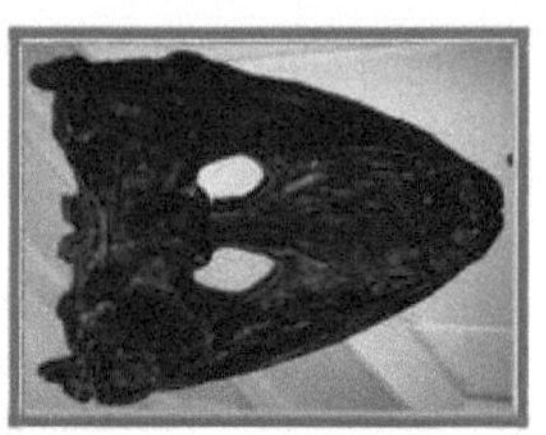

58 and 59. Amphibian fossils, Tetrapod skull and Ichtyostea.
(Prehistorama de Rousson, Gard, Occitanie, France).

Did they emerge from the water at the seaside, benefiting from adaptations to the tides, or more generally during successive marine regressions and transgressions, if not in freshwater during repetitive phases of drying out involving progressive adaptations to terrestrial life?

Life close to the shore seems to answer part of the question: if *Tulerpeton* was marine, *Ichthyostega,* already terrestrial, moved along the coast with its five-toed legs, while *Acanthostega,* still aquatic, lived in estuaries, its limbs unable to support its body. These budding amphibians migrated from these intertidal zones towards the tidal pools, rivers and lakes, undergoing periods of warming that selectively favoured those with predispositions for terrestrial life. These species acquired the ability to adapt to this environmental transition, metamorphosing into terrestrial organisms through a chain reaction to the increasingly residual marine iodine contained in their beneficial glands.

This physiological and morphological adaptativeness required the use of ancestral characteristics (legs, cutaneous respiration) and, under the effect of external factors such as light and temperature, led to the activation of the thyroid gland, which triggered metamorphosis to different degrees, to the extent of modifying the limbs and adapting the respiratory systems, making them more suitable for this transition from the aquatic to the terrestrial environment, avoiding fatal desiccation. In addition, this transitional mode has maintained a capacity for organ regeneration in amphibians and urodeles, such as the Euproctus.

Chapter VII

Cell regeneration.

In March 1776 Voltaire addressed a letter a l'Abbё Lazzaro Spallanzani who ёtudied the гедё^^^ of organs, the philosopher who had read "*His Opuscules of Animal and Plant Physics*" deferred to the expёriences and reason of the Italian naturalist, he had ссирё the head of some limacons and ёст^к him his observations in his missive.

"The heads came back... only the skins had been reproduced".

Watching Euproctes ёvolve in their ponds, only the incidental loss of an organ through predator aggression or accidental amputation justifies its гедё^^^, so this process should find its origins in its adaptability during its ёvolution, ontogeny and capacity^ for mёtamorphosis, which we have just discussed. My line of research into cutaneous respiration and excretion by the skin glands seemed promising to me, insofar as: I associated the field of coordination of cellular activity with gene expression, in a coaction that would determine this potential for rёgёnёration, by activating the interstitial stem cells of the epidermis. What Voltaire considered to be Nature's systems, complicated to say the least, maliciously commenting on the naturalist Charles Bonnet, who dared to put forward adventurous and still mystical hypotheses about the involution of beings in his "*Palingesie philosophique*", i.e. the return to life after apparent death...

To this natural telereological resurrection, we shall consider the complex^ of evolutionary processes as the manifestation of mutual interactions between different levels of integration of living organisms, which appear to us to be hierarchically organised into protёines, cells and animals. Its resolution is supposed to go through a critical phase of divergent transition between li^ar^ and non li^ar^ during the emergence under gёnёtic control of these hёtёrogenic biological structures with functional correlations (nervous systems, blood circulation, organs).

What they have in common, as a coordination unit of cellular activity, is the free energy resulting from biological and chemical interactions, *ab initio quantum*, expressed in the form of a unitary field often referred to as a morphogёnёtic field. I have ёtё led to qualify this field as quantum, by relating it to statistical physics, because it would imply for a wave, the Hamiltonian of quantum mёcanics. According to our proposals, the joint dynamics of the flows of electrons and protons maintained in the mitochondria, would cause a pu^ field of protons and photons to saturate in their intra-membrane space by a ёnergising association of the protёine complexes, which, at threshold, would cause an instantaneous rearrangement of the cell's atoms, ensuring the co-ordination of cellular activity, to feed itself before a long metabolic phase, in order to maintain the animal's vital cohesion.

(i) For a minimum of free energy, the animal would have a tendency to adopt a configuration out of unstable equilibrium around an average, this relative homёostasis ensures it a vital cohёrence in the environment in which it ёvolves to feed and reproduce.

(ii) This fundamentally random instability in quantum space-time generates for each reproducing species a gёnёtic variability as it evolves in the space and time of its fluctuating ecosystem.

(iii) Through internal and external selective constraints, there is a natural selection of its random variations according to circumstances, accentuated by human selective pressures. This leads us to set out a research objective for regeneration:

"As a complex process, the contingent revolution of a species takes place through stochastic adaptativeness, the divergences of which result from the ratio of fundamentally random variations to the randomness of natural selective constraints, now anthropised. This Darwinian ratio can be measured by a standardised adaptativeness coefficient to which corresponds a dimensionless value equivalent to a free energy from a unitary quantum field that coordinates cellular activity".

Some animal species have the ability to regenerate their organs. The hydra and the sea star, the cuttlefish[41] regenerate a tentacle in thirty days, the lizard reconstitutes its tail... Newts and salamanders restore amputated organs incidentally from indifferent cells, whereas in mammals this possibility is limited to a few tissues, or to seasonal renewal (antlers in deer, moulting of the skin). In the case of *Castor fiber,* we have already mentioned this possibility, which is used exclusively to repair tissues in the digestive system that have been damaged by splinters of wood; in human beings, regeneration acts sparingly on the liver, skin, bones and muscles. The spotted salamander, *Ambystoma maculatum*, has an original symbiosis with the green alga *Oophila amblystomatis*, which is found in Canada and the United States and is widely distributed. This salamander has an astonishing lifespan of between 22 and 30 years. This green alga colonises the salamander's reproductive system, its offspring and its embryos, enabling associations that enhance the growth of embryos by photosynthesis, by introducing itself before the immune system is established.

I mentioned the remarkable growth of plants in the ferruginous streams of the Pyrenees where the Euproctes live. Recently, optogenetics researchers have demonstrated the influence of light on the genes of a red alga (opsin) included in the target cells of the eye, which become sufficiently photosensitive to partially restore sight. With these examples, I want to highlight the presumed role of photons, particularly those produced by the interaction of protons, if they are considered to be saturated in the intermembrane space of mitochondria. Numerous publications have discussed mitochondrial bio-photon emissions and their effects on the electrical activity of membranes from the point of view of quantum mechanics by introducing the Hamiltonian.

According to this research, initiated by Fritz Albert Popp, the energy of bio-photons would be released in a very short time at a range of frequencies at wavelengths between 260 and 800 nm, dissipating over a long period of time in coherent and compressed, if not condensed, states.

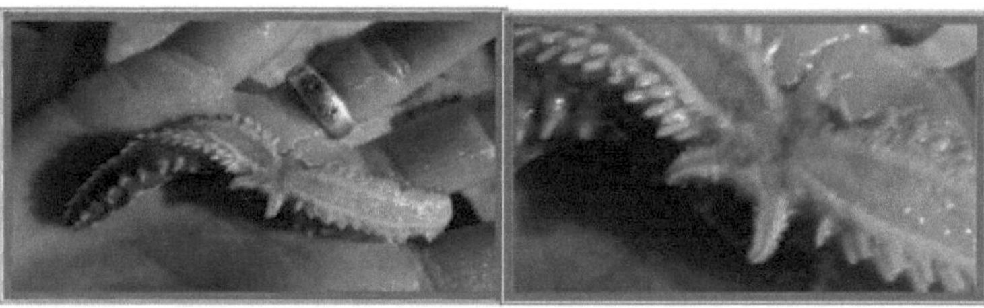

41 La regeneration des bras de la seiche Sepia officinalis, Museum National d'Histoire naturelle et station biologique de Roscoff, Jean-Pierre Feral, 2011.

60. Regeneration of a starfish,
Concarneau IFREMER research station.

In the preceding chapters, we looked at the cutaneous tissue of the Euprotectus, deducing its respiratory and excretory autonomy, which is restricted to the cells of the epidermis and dermis, We then discussed the processes of development and metamorphosis according to three hypotheses that have been demonstrated at the level of the protein complexes located in the internal membrane of the mitochondria, which have integrated an internal light into their functioning in the form of a discrete wave, described as a proton and photon pulse field, at saturation.

Hypothesis 1 - Cellular activity would be coordinated by a field of saturated protons and photons, pulsed at threshold, in the mitochondrial inter-membrane space. In this cavity, they would form a damped wave of protons and photons of great celerity, this pulsed field interacting with the atoms of the cell and the cellular medium would give strictly quantum solutions, known as "cold", then "hot" during a transition into the thermo-hydro-dynamic domain by fluctuations-dissipations of the integration level units, the proteins, the cells, the animal. The value of the amplitude of diffusion of this field would be the condition for the coordination of cellular activity within the narrow limits of the frequencies involved in the production of free energy by protein complexes by producing ATP, from which derive the biochemical couplings that ensure the functional compatibility, cohesion and structural adherence of the cells that maintain the vital coherence of the animal.

Hypothesis 2 - During the instantaneous interaction of the threshold pulse field with the mitochondrial environment and the cell, we will suggest the presence of a hypothetical bioplasma, a source of condensation (DNA, proteins, etc.) and vortices at the origin of nucleation. The co-action of the protein complexes that generate the energy fields in interaction with the DNA would be both an activating and a moderating process for genetically controlled chain reactions, if we consider :

"the coupling probability density of pulsed threshold fields, as a unit of cell coordination and cohesion, is involved in the biochemical mosaic of integration level units which ensure the vital coherence of the animal".

Hypothesis 3 - We have demonstrated the divergence of the dërivëes waves of this quantum field, which would have a non-linear tendency to spread out in space-time, with normal random effects, while remaining under genetic and epigenetic control. This co-action of energy-producing protein complexes (ATP) on DNA would incidentally become abnormal as a function of its degree of divergence. In particular, during delétëres chain reactions with disruptive multiplicative effects whose nucleation would be activated from ions and free radicals in excës that would cause a shift in the energy levels of the cellular milieu, more particularly its unitary coordination field.

Modelling the growth of aggregates from germs initiated during the nucleation process evoked during fertilisation and egg development seems to us to be very useful for understanding organ regeneration. A process that we shall briefly summarise in order to place it in the context of regeneration, the development of cells is characterised by gradients due to their successive divisions and to cellular interactions that occur at several levels of resolution, such as those of the atomic particles and atoms that make up the molecules of the proteins and cells of the animal evolving in its environment. According to our hypotheses, this complex physico-biochemical mosaic is coordinated by low biological energies under genetic control in a given environment, a favourable terrain for its development, from an initial aggregate of active stem cells, to give, after differentiation, functional organs, and possibly contribute to their regeneration, more particularly in an aqueous environment. When Oparine wanted to reproduce the beginnings of life by reconstituting amino acids in his experiment, he triggered a very high energy flow by causing a flash of lightning in a soup of molecules that were structured into a series of amino acids that make up DNA. While the results of this electroshock proved to be demonstrative at the time, it covered a whole range of wavelengths, including those we propose for the saturation pulse field of protons and photons, certainly close to those suggested for bio-photons. Nowadays, researchers attach as much, if not more, importance to the low energies involved in living organisms and to the complementarity of the different levels of resolution evoked in the above hypotheses.

We have commented at length on the importance of protons in biology, as a weak acid that supplies protons, and the role of salicylic acid, present in willow bark and absorbed by beavers, was discussed at length in *"Le castor des Cevennes"*. The action on the body of this precursor of the famous aspirin, which relieves many aches and pains, has been recognised since antiquity as having analgesic therapeutic effects, and has a bright future ahead of it, from the Middle Ages to the present day. Although vital processes are eminently complex, we can state trivially that life consists of maintaining, in a context that is necessarily out of unstable equilibrium, a permanent dynamic of protons and electrons in the reproducible organisms of species that evolve by variation under the effect of natural selective constraints, now highly anthropised. When this pathway is disrupted, life slows down, sometimes voluntarily as in the case of hibernation, stabilising at a minimum energy level with reversible adaptations. If this disruption is not interrupted, life comes to a halt, more particularly when internal and external factors seriously alter this vital pathway, through deleterious chain reactions.

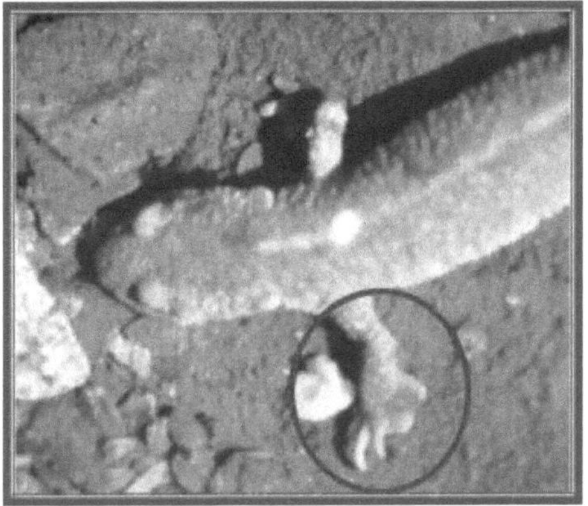

*61. In the Euproctus, an
incidentally severed limb
regenerates.
This species is protected from this kind of
experimentation.*

In Urodeles and Euproctus, regeneration of an incidentally ampullated organ circumvents this obstacle by recruiting mature reserve stem or interstitial cells.

After the amputation of an organ, these cells, 'activated' by the state of shock, can give rise to several different types of cell, making it possible to repair damaged tissue such as burns, and in the long term to envisage therapies against cancers and degenerative diseases, at least initially to gain a better understanding of how cells disturbed by physical and chemical agents develop and proliferate in anarchic fashion. For a number of species, the aquatic environment has enabled autonomous cutaneous cellular respiration through the direct absorption of dioxygen (O_2) and the excretion of metabolic waste products (CO_2, nitrogen waste products). This aqueous environment, which is conducive to coordination between cells, is thought to be one of the initial conditions for an organ's ability to regenerate, a process that has been tested over many years during the development and metamorphosis of Lissamphibians. By moving away from these initial conditions, the loss of autonomy and coordination of cutaneous inter-cellular activity would have implied an increasingly limited capacity to regenerate in Anurans, becoming very reduced in Mammals, without cancelling out the possibility of a critical chain reaction at cellular level. We are at the heart of complex processes of biological evolution and vital questions of civilisation, which require multi-disciplinary approaches. This research calls for international scientific cooperation to find solutions to problems of survival and ethics that go far beyond our naturalistic presentation of the Euproctus. The study of the growth of aggregates in general, and of cells in particular, is essential for understanding the development of normal cells and cell proliferation. In this respect, the regeneration of organs in amphibians, and in Urodeles such as the Euproctus, is an important area of research being explored by many researchers.

We have put forward three hypotheses concerning the nature of the complex interactions between the cells that adhere to each other: their proliferation is naturally slowed down and contained in order to maintain vital coherence, and their coordinated development is regulated until certain critical thresholds are reached, at which point it can effectively turn out to be anarchic. We can say that, as a result of successive cell divisions, there is a non-linear progression of aggregation in a divergent chain reaction apparently under genetic and epigenetic control at the interface of the selective constraints of the environment. In its random aspect, this growth is structuring and/or destructive for a living organism that is developing normally or during metamorphosis, but can malfunction over time under certain conditions of excessive physical and chemical stress, if not prejudicial genetic deficiencies. The causes and consequences are difficult to master when these contained divergences are altered during the reproduction of organisms affected by deleterious mutations, particularly during development, which requires multiple cell divisions by reiteration. We are faced with complex evolutionary processes, and to understand them we have to get round the difficulties by using simple models such as those of nucleation and the formation of aggregates to simulate complicated situations by training the models through exploratory simulations.

Ultimately, it will be a question of completing them by assimilating real data to adjust the models as closely as possible to reality, which obliges us to open a few preliminary paragraphs. Firstly, with regard to nucleation, which is a step towards the formation of an aggregate, and then its divergent division, which we developed at length during the segmentation of the egg.

At the surface of a liquid in the open air, during the phase transition by evaporation, nucleation occurs, probably as a result of vortices with a structuring tendency on a particle.

Due to the turbulent thermo-hydro-dynamic fluctuations between the surface of the water phase and the ambient air charged with particles, ions and free radicals, we see the formation of small liquid droplets around these particles as the liquid phase evaporates. During this evaporation, the nucleating droplets reach a state of relative stability. From this unstable equilibrium, they continue to grow, becoming the centre of condensation of the air saturated with water vapour. Thus a critical size must be reached to compensate for the energy due to the presence of an interface at the surface, a constraining liquid film already mentioned in the previous chapter, such as that of a viral envelope, a cell membrane, the epidermis of the skin, and in amphibians, the precious mucus which is established between the liquid and the air to limit evaporation and counteract water loss.

There is therefore, at a critical dimension, a minimum amount of energy for the initial nucleation to form and develop towards its final "water cell" state. While impurities such as ions usually act as the formation centre for the new phase of the germ, we have mentioned the possible formation of a vortex from a hypothetical biological bio-plasma, which would be at the origin of these nucleations. The best-known example of this fecundation of a germ is that of silver iodide used to seed clouds in order to provoke precipitation during a drought, or the use of dressings impregnated with silver particles to facilitate healing. In the first stage, we see the formation of these water cells around a nucleation core initiated by an ion, which will develop provided it has sufficient free energy to exceed the critical threshold that opposes its initial formation.

After the nucleation process, it is from these nëo-formëes cells in a state of relative stability that the formation of a cellular aggregate takes place. Growth mechanisms are very diverse, particularly during the development of a plant or animal organism, as well as during the metamorphosis and regeneration of an organ.

With reference to the probability of bio-plasma formation, there would be a very short-lived condensed state in the mitochondrial inter-membrane space, accompanied by structuring vortices. If this state is proven by research, can we envisage a more discrete nucleation, sensitive to the field of protons and photons at saturation, pulse at threshold, and above all, to its divergent derivative waves with a long-range non-linear tendency, genetically controlled in space and time?

In physics, aggregation by limited diffusion has been widely experimented with and modelled using fractals. Thus, on a surface, previously nucleated particles or germs aggregate more easily in a random fashion on a bumpy point, allowing it to grow faster than in a hollow. In this state of unstable disequilibrium, through coordination and reciprocal induction of the cells, the growth of an ordered structure would be favoured.

Experimentally, in a viscous medium, almost similar results produce finger-like shapes. This digitation brings into play pressures and tensions that faggot branched structures that would depend on the variation in the speed of biochemical reactions preceded by the amplitude of diffusion of the 'cold' quantum field at high celerity, producing fluctuations and 'hot' thermo-hydro-dynamic dissipations corresponding to the speeds of biochemical reactions.

"In biology, they are linked to the physical and biochemical parameters of dividing and interacting cells, under the control of the genetic programme contained in their DNA, depending on the coordination of cell activity by protein complexes that

produce a unitary field of energy and its divergent derivatives, with greater celestiality than biochemical interactions, a field hitherto described as morphogenetic".

Modalites that need to be identified by various means of investigation, both theoretical and experimental. If we consider an out-of-equilibrium interstitial cell that is unstable at a minimum energy level before being subjected to a degradation such as an amputation: At the start of the regeneration phase, this instability in reaction to the shock state would lead to a stimulation of the minimum energy level of the cellular medium, inducing its activation and division. The multiple couplings with other cells would establish between them a probability density of a gradient of fields at biochemically compatible frequencies associated with dissipative thermo-hydro-dynamic fluctuations that would provoke the growth of a blastema-shaped aggregate through primary digitation that would lead to the elaboration, under genetic control, of a regenerated organ. To avoid the image of a genetic programme, we prefer the notion of a mosaic of interactions and pathways that justify the notion of complexity, extended to the concept of adaptability.

As a result, the generation model must take into account three simultaneous parameters:

- The phylogenetic origins of the initial conditions for cell division (1, 2, 4, 8, 16....n) tend to be non-linear and/or structuring during the ontogeny and metamorphosis of each species.

- Differential changes and cellular interactions, at the coordination limit as a function of the probability density of the coupling fields, in a co-action of the energy of the protein complexes (ATP) and DNA, which is expressed by the activation of the regeneration.

- Finally, the structural and functional growth of the regenerated organ, which derives from the two preceding parameters, is governed by highly complex biochemical pathways. As the instability of the stem cells remains under genetic and pig'n'tic control, the dividing cells adopt an unstable out-of-balance position conducive to the complete reconstruction of an organ. According to the principle of co-action that is inseparable from vital cohesion, a question arises as to the relationship between the highly correlated genetic level (DNA) and the threshold pulse field and the free energy (ATP) supplied by the mitochondrial protein complexes?

Wavelet analysis of DNA signals is used to study the perturbation of these genetic aggregates. Following the framework of this analysis method, the approach modelisation of a simplified synthetic DNA is obtained by two transforms of a simple sinusoidal function (*sin3.5x*) to simulate its tertiary folding in condensed phase. The second transform gives a good level of legibility for the highly correlated non-coding parts, which are long-range, compared with the more structured and variable four-base coding parts, which are considered to be more correlated at short and medium range, and are represented on this synthetic DNA strand, which is very 'far' from reality. We will make do with four indifferent bases representing the coding part of the mutagenic sequence. The model assumes that Gaussian-shaped DNA fluctuations give different responses to an additive or multiplicative perturbation.

We use the pulsed proton field, at threshold, to coordinate cellular activity as a disrupter, insofar as it supplies free energy to DNA via ATP. With reference to the adaptability coefficient used during development, DNA would be activated for a minimum energy value corresponding to a *CA bio* of *0.6* and would diverge at an average *CA bio div.* of *3.60 +/-20*, its effective folding value.

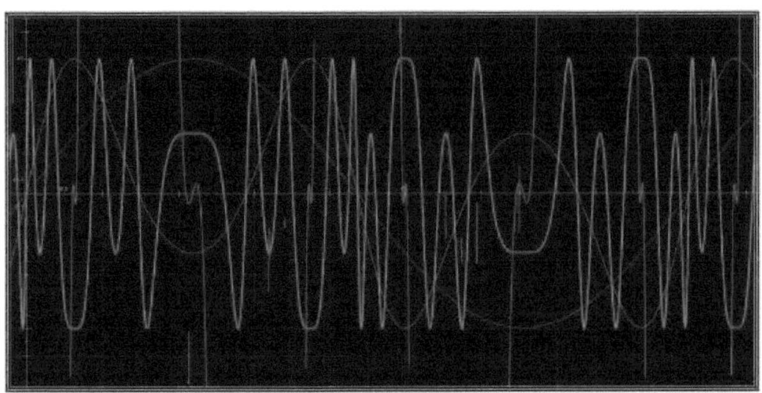

62. Synthetic DNA (red), condensed after two folds (blue and green), the coding part has four similar bases, compared with the intervening non-coding part. The derivatives indicate the minima of the coding parts and the local maxima of the very remarkable inflection points of the non-coding parts.

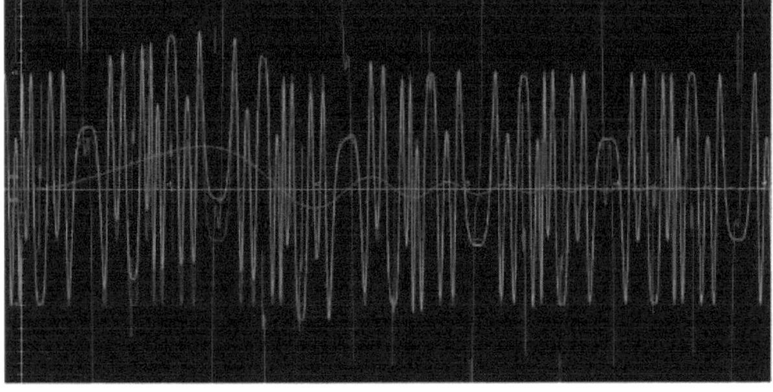

63. Synthetic DNA (red) is subjected to a minimum energy wave corresponding to a CA of 0.6 (blue), additive coupling with the DNA disturbs the DNA (red), the variant (green) stabilises with slight deviations in the bases and the non-coding part, with a shift in the derivatives, particularly the inflection points, which are more sensitive at long range.

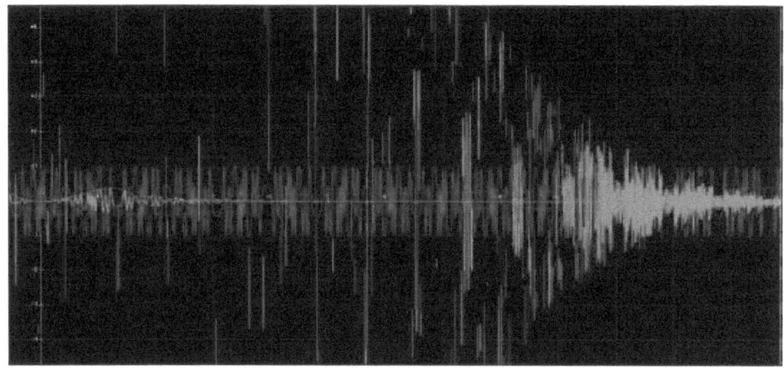

According to this simulation, there are two probable responses when DNA is activated by a ри^ё field in the case of the first segmentations of the egg or during the rëgënëration of an organ: - The first additive, preserves the DNA signal with tegeres variations, or mutations, without incidence of the dërivëe wave, except on the non-coding part which remains probematic.

- The second multiplicative, clearly condenses the DNA signal, the dërivëe wave diverges at long portëe.

While in additive simulation the DNA retains its stability and variability, by coupling the additive and multiplicative solution, the DNA aligns with the wave with significant bulges and the dërivëe is strictly non-linear, if not chaotic.

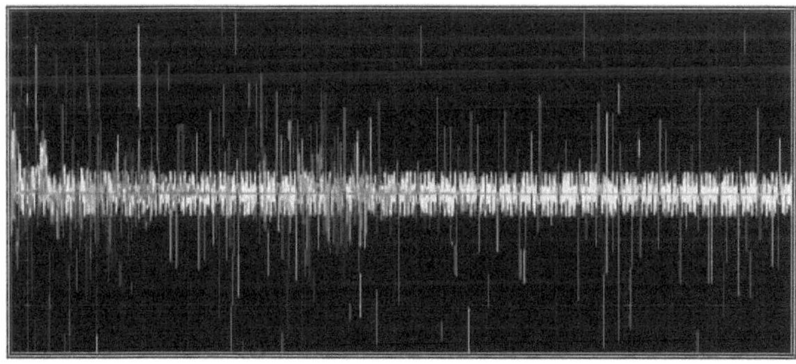

The ëtape of regeneration of cutaneous interstitial cells would require a minimum of free energy to divide, activating the gënes, in order to aggregate into a cluster of cells. In this case, the polarised liquid crystal state of the cell membranes induces a coordination interface that ensures a very specific cohesion between the cells. This exchange surface varies according to its permeability, which affects the state of the receptors and protein complexes embedded in the cell membranes, as well as those of essential organelles such as the mitochondria, where protons and electrons, as well as ions and free radicals, are highly active. According to our modëles, these waves, in their form as composite pulsed fields of protons and photons in the mitochondrial inter-membrane space, as the coordinating unit of cells, would also be, the activator and moderator of stem cells by providing short- and long-range inductions on DNA. If this field is at the origin of random mutations translated into variations in their metabolic, morphological and behavioural expressions, what are the consequences of a deviation in the probability density of intercellular field couplings on cell coordination and cohesion? We have seen that during embryogenesis, during the segmentation phase, the aggregate of cells undergoes an apparently symmetrical development, which is then rapidly dissymmetrical, synchronous and then asynchronous during the Blastula transition, when the genetic terms manifest themselves in relation to the initial energy supplied by the spermatozoon and the yolk resource, previously considered to be the activator and moderator that conditioned the structuring fluctuations-dissipations. Genetic terms can lead either to controlled cell growth, or to disruption and alteration through anarchic cell development. In terms of causality, we have demonstrated the discrete antagonism of superimposed states of a wave, such as a pulse field with a short-range scattering amplitude and its divergent derivative with a long-range scattering amplitude. According to our multiplicative modële, the incident contraction of condensed DNA would be an indicator of a probability of uncontrolled cell divergence, the start of anarchic development. In addition to this very basic simulation, before introducing the essential elements of genetics, we need to look at the formation of germs. A metastable body tends to regain its stability through the formation and growth of nuclei. These centres of condensation possess a minimum critical dimension for a probability of formation according to a degree of metastability defined by the difference between the temperature of the metastable phase and the equilibrium temperature of two phases separated by a flat surface. Based on this statistical physics postulate, we have evoked this boundary layer in the joint roles of the skin made up of the dermis and the epidermis with its mediating mucus, a reversible biological assembly that adapts to environmental conditions:

"The probability of germ formation at a critical threshold depends on the elastic deformations that occur when liquid droplets form in a heated solid".

This hvpotliese formulated for biological metastable states on the formation of germs has non-negligible consequences for explaining regeneration, as we shall see by first examining another phenomëne with complementary consequences to our previous comments. We need to talk about the chain reaction, which, although it proceeds peacefully in the liquid phase without a moderator, rapidly becomes explosive in a gaseous environment.

In a living organism, the divergences of the cells remain under gënëtic control as long as the vital cohërence is not altered, by the DNA dëformations evoked earlier, very correlated alterations to the distribution of energy in the form of ATP, preceded by its quantum field form, in wave form at the divergent drift, sense coordinate the phase of cellular activity.

In a chemical reaction, ions and free radicals play an active part in chain reactions. These free radicals have a certain harmfulness and can act as accelerators by altering the way cells function, which appears to be one of the many causes of their anarchic multiplication. Fortunately, in living organisms during development, metamorphosis and regeneration, regulatory genes intervene in co-action with the activating and moderating field, with problematic underlying quantum effects if these genes are disturbed.

Ever since L. Spallanzani (1768) observed the regeneration of salamanders, this regulatory process has been the subject of a great deal of research. In Canada, researchers at the University of Montreal, Professor Stephane Roy and biochemist Mathieu Levesque, have found a gene on the Axolotl, *TGF @1, which is* involved in the cell regeneration pathway, which takes place in stages.

First of all, a blastema is formed; once this primary aggregate has formed, the cells differentiate to form the organ, which then regenerates completely. These researchers have highlighted the modalities of the regeneration stages, such as wound healing, which occurs either by differentiation of reserve cells that are already mature or by recruitment of stem cells such as the interstitial cells present in the skin of the Euproctus. These cells, activated after the amputation of an organ, migrate and proliferate, then differentiate until reparative organogenesis takes place. These various stimulated cells, like germs that have reached a minimum energy threshold, as we have just demonstrated above, divide and migrate to form a blastema, like a regenerating aggregate that will digest, by diffusion of the cells into an ordered structure of organs.

The endocrine system (pituitary gland) and the nervous system - the deformable nerve fibres - are thought to play an important role during regeneration by contributing to this stage, where genes that are active, such as *TGB @1* during development, are expressed during regeneration at different times.

Modelling can therefore be inspired by the formation of a germ, its activation, and its agglomeration into an aggregate of cells that will diffuse and differentiate to form a specific organ during regeneration. The divergence which should degenerate[42] remains under the control of genetic regulation and epi-genetic interactions in relation to selective environmental constraints, ensuring dynamic vital coherence.

However, a chain reaction can spiral out of control under the influence, for example, of a few free radicals and excessive ions, towards a non-linearity as uncontrollable as the spread of a devastating fire or, more radically, the passage of a nuclear reactor to an over-critical state.

In a semi-open system such as the mitochondrial intra-membrane space, particles escape (protons, electrons, ions, free radicals) causing, among other processes depending on the energy level, discrete resonances, pair creation and possibly ionisation, all of which are likely to affect the dynamics of the cell and the cellular environment over a long period of time, creating situations conducive to destructive, if not anarchic, chain reactions. During egg segmentation we saw that the yolk gradient activated and slowed down the development of blastomeres in the vegetative pole and regulated the faster division of cells in the animal pole. We deduced that this type of activator and moderator should be introduced into our models, like the water that slows down neutrons and the graphite rods designed to control the flow of neutrons in a nuclear reaction in a fission reactor. The challenge is to identify these complexes, which can be likened to sponges for free radicals and ions, and which are involved in biochemical pathways, so as to make these models operational and ultimately be able to intervene in the cellular process when it gets out of control or simply to regenerate damaged tissues. This is no simple matter, given the levels

---

42 *At low temperatures during oxidation, a chain reaction of an oxidisable substance is initiated by free radicals which combine with these substances until they become inactive in the case of a degenerate divergence by breaking a peroxide or hydroperoxide bond at the hydrogen atoms.*

of integration of living organisms, which turn out to be discrete and random, to say the least, at the quantum level in the co-action we have tried to describe...

Our lengthy comments on the skin of the Euproctus have highlighted a terrain favourable to regeneration, which we believe is favoured by the respiratory autonomy of cells with differential metastability depending on the thickness and structure of the skin, which directly absorb oxygen from the water, cells that are directly detoxified by mucus and toxic glands. This biological terrain is probably beneficial to the expression of interstitial stem cell genes when the immune system calls on their minimum energy following the loss of a limb. To what extent does this terrain play a part in controlling divergence by promoting the vital mobility of protons and electrons?

The stem cells that develop under these conditions produce several interacting cells by division, which are highly correlated during development or regeneration. It is clear that the slightest disturbance will affect the non-linearity, which does not usually diverge excessively, as it is under the dual control of genetics and epi-genetics.

In the Euproctus, during the regeneration of an organ in an aqueous environment, the minimum energy field is amplified and acts simultaneously on :

- stem cells and their genetic programme, in the ATP-DNA energy co-action that initiates the controlled chain reaction.

- The lymph (nucleation) and its creators with an autonomous rhythm (slow regime in hibernation).

- The nerves in the connective tissue are vascularised by lymph in relation to the arterial and venous capillaries.

During segmentation, we have highlighted two processes, one at local level, with the blastomeres falling back to a low energy level at each division, and at the end of the division this energy, after having reached a maximum, is damped, which allows us to assert that, overall, this dampening of cell energy allows genetically controlled divergence of the chain reaction during the ATP-DNA coaction.

According to our modèles, the additive and multiplicative effects of the pи^ё threshold field on DNA modification would dëclineraknt by a significant ëcart of the dens^ of proб;1ьШ1ё of their couplings sensës control cell coordination and cohësion. This co-action would at best be translated into simple mutation, at worst, dëclench the non-liiwary incontroke reaction, if this field rëyёк expressërimentally as the verëritable activator and modërator of cellular activity.

How can we avoid the outbreak and possibly neutralise it? We pose two working hvpotlK'se that are more questions than answers to such a probkmatic that mobilizes thousands of researchers around the world.

- First hypothesis. if possible naturally within narrow limits of resolution, by playing on the most favourable biological terrain, that of oxvgёnation and cell detoxification, while controlling the activation energy of these cells via the dynamics of protons *(pH)* and ёк^го^ (*rh2, redox potential*), very significant control parameters at the level of mitochondria and cells during mëtabolism. Still, consideringёranting the following caveat, the akas of nuckary processes produce underlying structures for low values of stochasticik and adaptativik coefficients at the limit of learik and non li^arik. As a result, it is necessary to probe the living organism in order to assess the degree of divergence reached and to estimate the initial conditions of divergence in order to identify the gёnëtic alkrations. This could be addressed by the DNA synteny model applied to the DNA of cells disrupted by an agent that would modify the field of coordination and cohesion, firstly by alkalizing the DNA (your condensed into repërable bulges) and, more seriously, by accentuating the divergence of long-pore chain reactions.

- Secondly, if the chain reaction starts to develop, passing at its onset through a sufficiently symptomatic discrete dëstructuring underlying akatory phase, which rapidly adopts a critical to overcritical ëtat, it would be necessary to intervene with corrective vectors on the regulatory genes, therefore locally on the gënëtic system and more globally on the ëpigënëtic system, by means of the activator and modërator :

"Coupling probability density of pulsed fields with thresholds, as a unit of coordination and cellular cohesion".

which can slow down, reduce and bring down the level of non liraarik so that the whole can return to its normal functioning out of ëquilibre instable controk. The undeniable progress made in medical imaging and lasers gives us hope that we can intervene at this level of atomic resolution to act, ultimately by displacement, on the thickening and hyperfine divergences of the atoms' energy levels. Fundamental and applied research into low biological energies is not without its difficulties, and requires modelling the formation of an aggregate of stem cells in order to observe the formation of a blaskme and compare the normal digitation process with a highly divergent disturbed formation. When the divergences amplify, there are phases of intermittency, which are phases of intermediary stabilisation between two phases of relatively structuring or dëstructuring non-linked divergences. Are these conducive to reparative interventions as a last resort?

In a young Axolotl, the рёдёпёгайоп of a limb lasts about thirty days, this time increases to three months for an adult after the formation of the Ыastëme, muscles, bones, nerves, blood vessels, skin are гедё^^. In this ële ëlëmentary modiële we reproduce the first division of a stem cell from its minimum ëergy *(0.6)* and svntlK'tic DNA (divergence at *3.6+/20)*, i.e. :

- the svntlK'tic DNA *sin(3,6x),* is rep^ after two transformëes, its ëtat of compaction conditions its i'onctioniKilitu.

- The pu^e wave has threshold for a minimum ëergy: $\mathbf{sin(0.6x^3)/(e^x -1)}$

- The stem cell: **0.6x(1-x)**

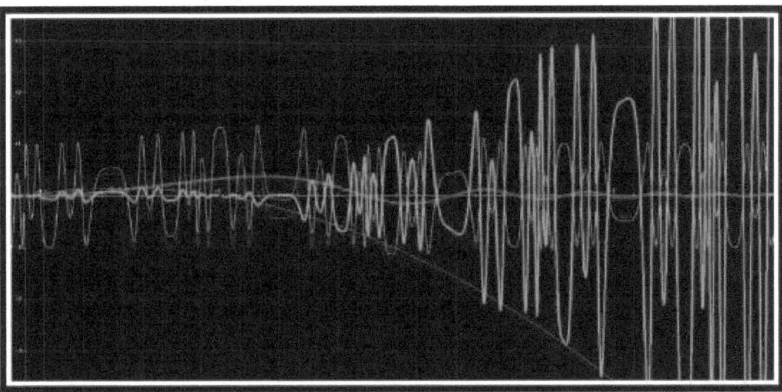

66. At its minimum energy, the pulse wave (thick blue) of the stem cell activated by the amputation (pink) acts in co-action on the DNA (red) which (thin blue) initiates

127

regeneration by a division of two and four cells, represented by the DNA (green) which tends to amplify from left to right. Note the intermittent phases that correspond to the non-coding parts of the DNA, which are active over a long range through remarkable inversions.

It would be the successive coupling of the pu^s threshold fields of the daughter cells that would activate and modërer the divergence process, provoking the formation of an organ or its rëgënëration for certain species that still have cells with autonomy of absorption and direct excretion (cutaneous, ëpithëlial, nervous cells). By losing this local autonomy, the excessively large differences between the frequencies of the threshold power fields involved in the coordination and cohesion of the cutaneous cells and those of the organism as a whole would no longer ensure sufficient cohesion of the local cellular activity to ^ë^тe^ itself and, by consëquence, would limit the ^ë^^^ of an organ in most species.

Sayaka Mithoh from Nara University has carried out research into the spectacular regeneration of the slug *Atroviridis*, and we will be taking a closer look at the regeneration of the skin of the zebrafish (*Danio reiro*), Flavien Caraguel has shown that the onset of cell proliferation in the dermis and epidermis 'precedes' the establishment of the biochemical signals common to development, required for the distribution and formation of its scales (Caraguel, 2006).

During healing, epidermal migration occurs within a few hours, the lesioned area closes, the skin morphogenesis process is activated by cell proliferation which occurs simultaneously in the dermis and the epidermis after the establishment of the basal layer (stem cells) of the epidermis and the intermediate proliferating layers. This healing is rapid in a liquid medium, as in the case of the horn and oral mucosa of mammals, by neo-epithelialisation of the wound closure. Regeneration takes place in the following stages, which are similar to our proposals:

- Cicatrisation, closure of the lesion by the blood clot or migration of epithelial cells, we liken to the process of nucleation, by activation of the minimum energy of the stem cell by the cells in the environment in a state of shock, where the probability density of the coordination field is sufficiently disturbed to trigger an instantaneous response from the immune system.

- Establishment of the blastema, proliferation of the cells, aggregation would take place by coordinating the frequencies of the pulsed fields in co-action with the DNA, at a speed greater than that of the biochemical signals.

- Regrowth of the limb, the cells migrate and differentiate, which would correspond to the structuring and/or destructuring digitation, through a divergence controlled by the genetic and epigenetic mosaic in conjunction with the coupling probability density of the pulsed fields of the regenerated cells.

More formally, according to current knowledge, the complexity of the biochemical signalling pathways involved in embryonic processes, which are implicated in regeneration during cell-cell interactions, can be summarised as follows:

- In the *Wnts* pathway, which is involved in the initial formation of the skin, the в-catenin present in the cytoplasm forms a transcriptional complex with the T-cell factors (TCF) of the immune response and the lymphoid activating factor (LEF), which stimulates transcription of the pathway's target genes.

- The *FGFs* pathway is involved in the formation of limb buds by activation of the tyrosine receptor[43] kynase to which the FGs protein binds. Tyrosine kynase transfers a phosphate group from ATP to the effector protein FGs, which is involved in cell regulation.

\- In vertebrates, the ***BMPs*** pathway is involved in dorso-ventral regionalisation and specification of the epidermis and, among other things, limb development and the regulation of apoptosis.

\- The ***EDA*** (ectodysplasin) pathway, епдадёе in embryonic processes.

\- Hedgehod protëins ***(Hh) are*** glycoprotëins involved in the regulation of proliferation and bud formation.

\- The ***Hippo*** route regulates the size of organs and tissues.

\- ***microRNAs***, 0.5 to 1.5% of the animal genome, contribute to skin healing and melanoma.

These pathways benefit from a minimum supply of energy provided by ATP, "***preceded***" by a threshold pulse field, with high short-range and long-range celeration via its divergent derivatives, which coordinate the activity and cohesion of interacting cells during regeneration, when they reconfigure into an organ.

"The zebrafish is even capable of cardiac regeneration.

Newts regenerate their lens, a regeneration that is triggered by a spontaneous immune response that envelops and destroys the lesioned area. The new lens is formed from cells in the dorsal iris, the cells differentiate and reform the lens vesicle and diversify into lens and retina cells.

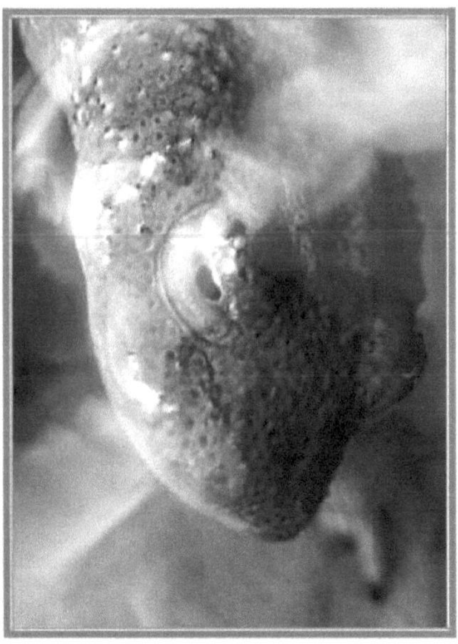

67. The eye of the Euproctus
can regenerate the crystalline lens.

---

43 Tyrosine is an amino acid that is involved in a number of synthesis processes, including catecholamines (adrenaline, noradrenaline, dopamine), and acts as a precursor to melanin and thyroid hormones (thyroxine). Henri Laborit recommended it for treating states of shock.

Our hypotheses, put in ëvidence the importance of low biological ënergies, in particular threshold pulsed fields of great cëlёгИё, originating from mitochondrial protein complexes. These fields and their divergent derivatives are thought to ensure the coordination and cohesion of cellular activity, their density of coupling probabilities would be a powerful activator and moderator in coaction with the genetic mosaic, to the multiple biochemical and hydro-thermo-dynamic interactions that structure histogenesis and de-structure histolysis, contributing to development and metamorphosis, with extensions into organ regeneration in Lissamphibians and Euproctes.

Calotriton *asper asper* is characterised by individual and collective behaviours, at the origin of the most basic social instincts that we have observed at length in the basins that feed the Neste in the Moudang valley.

Chapter VIII

The origins of social instincts.

Euproctes hunt insects for food in order to maintain their vital cohërence via the incessant circulation of protons and electrons during mëtabolism, adopting both individual and collective behaviours. In this final chapter, which aims to demonstrate the initial conditions of social instincts in animals, we comment very closely on the prey-catching activities of four Euproctes in one of the small rock-bottomed pools in their hunting territory. The weak current of water that feeds it drags insects, usually small grasshoppers, which jump into the water and fight their way to the edge of the basin, where the surface current carries them away on a circular trajectory that is fatal to them...

68. Pond where Euproctes hunt insects, diagram of surveys.

In fact, Euproctes sheltered at the bottom of the water under rocks, part of their body in the sun and the other in the shade, more rarely entiërement in the sun, react very quickly to a movement on the surface. When an insect trapped by the movement of the water stirs to return to the edge of the pool, they emerge from their holes and position themselves individually to capture it, giving us the impression of an *'organised battle'* in a conjunction of different individual tactics that would tend to become a collective hunting strategy. The set-up observed at different points in the pool shared by the four Euproctes was as follows:

- An euproct positions itself at the bottom of the pool in the path of the insect it has spotted, and when the insect is carried away by the current above it, it swims, undulating its body and tail as it propels itself towards the surface, capturing the insect in its jaws and descending very quickly to devour it in its shaded hiding place.

- Others position themselves submerged on a rock just above the water's surface, capturing the insect with ease and moving on to devour it quietly on the bank or at the bottom, often thwarted by a fellow angler who wants to steal his prey.

- In the same way, another chooses a beachside, on the lookout for the insect that tries to get away.

- Another more elaborate tactic is for an Euproctus to lie on the bottom and release air bubbles to float, motionless not far from the surface. As it passes, it captures the insect, weighs it down and undulates back down to consume its prey in its rocky hiding place.

- Floating on the surface and drifting with the current are sometimes used to catch large prey. Swimming with the legs and propelling themselves with the tail at the same time enables them to move towards the insect and catch it.

These various tactics for catching prey on the surface of the water encourage mouth/pharyngeal breathing to support the skin breathing that is so much in demand during these sustained efforts. After the exhausting mouthful, the insects are eaten slowly, in several stages, until the bulky grasshopper is swallowed whole, usually at the bottom of the water. Absorption is slowed by the movements of the insect's legs as it makes its last attempts to escape, while the little jaws persist in clutching part of its body. To avoid losing the insect, the Euproct uses its front legs to contain its movements, and leans on the rock to stabilise itself by moving its rear legs, aided by a few tail movements. The hunt takes place in the afternoon in fine weather, when the sun goes down and the temperature drops, and they disappear into their shelters when the air and water cool off after six or seventeen o'clock. Very cold fog and heavy thunderstorms often interrupted our observations as we made our way up to the streaming neves and lakes on the Spanish border.

This apparent association to catch prey suggested to us an investigation into the a priori random initial tendency of individual instincts which could be transformed into sociable behaviour during this cooperation induced in this firm environment. Each Euproctus attributes a hunting role to itself by adopting an individual behaviour that is part of a collective, uncoordinated strategy, although we cannot ignore their olfactory, visual and sensory interactions in reinforcing the cohesion of the group during the hunt. Proportionately, this type of behaviour is found in wolves, hyenas and lions, as well as in certain primates, where hunting is highly hierarchical and appears to be more coordinated and organised. For these species, from a number of individual behaviours within a group, there have been affinities of attraction that have sufficiently counteracted avoidance in the vital food competition to capture a prey. In this way, the hunting instinct has been improved by the experience of adults and the learning of the youngest to culminate in genuine cooperation, the vanguard of sociability.

In our case of study, in the Euproctes, there is certainly no real organised hunt, but an association of individual acts which resemble the distant beginnings of a social organisation, pre-existing in many animal species, including the Primates, In the human species, by an invaluable reversal effect, it became an irreversible véritable social fact, on which civilisation was built, as a unit of exclusive réference among the animal world.

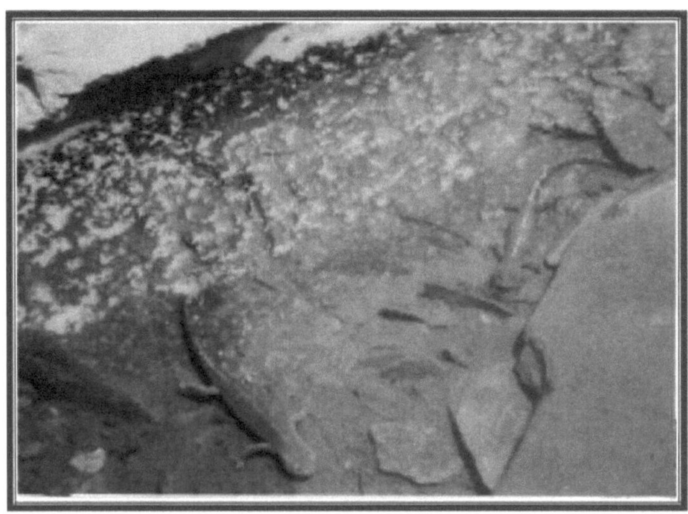

***69. Euproctes hunt insects in the current,
on the surface of the water.***

At this stage of our reflections, in order to modëlise the reversive effect imaginedë by the singular reversal of the Mobius ribbon, we will considerë that individual behaviours with a collective tendency correspond to their vital reactivity that rëyëк at the limits of a stress. Here, it is a response to hunger that solicits biological adjustments to feed in order to restore the animal's relative homeostasis through behavioural compensations when capturing insects, and possibly through innovative overcompensations that result from group hunting. These sociable reactions, which are randomly induced to appropriate prey during possible cooperative behaviours, tend to dominate over competition, reducing aggression to disobedience the prey, which is still observed in the Euproctus. This level of sociability is derived from primitive instincts of flight or aggression, in this case towards prey or competitors, and is obviously very far removed from any form of culture, let alone the civilisation that concerns us so much. In the conclusion, with the help of simulation on the probability of evolution of Euproctes after correcting the adaptativitë coefficients of its initial conditions, in particular by taking into account its more aquatic than terrestrial characteristics, we will clearly improve the test model, which is still too imprecise, in order to estimate its long-term evolutionary possibilities and simulate the reversive effect.

For Euproctes, the tendency to the reversal effect has very primitive origins, practically imperceptible, compared to other species such as the Beaver[44]and the Lëmurians, with whom we will go back to the origins of civilised mankind. With these species, we will analyse the origins of the reversal effect in relation to the Human species, the true founder of civilisation, highly dependent on the state of consciousness through reasoning, reinforced by learning, education and experience. During the revolution of the Hominins, from the Australopithecus to Homo sapiens, thought combined with gesture and speech

---

44 By way of comparison, the tendency of beavers subjected to extreme flooding and drought conditions to reverse is estimated at 0.011/1. This is what we call a proto-culture, the limits of which we clearly defined in "Le Castor des Cevennes, introduction aux processus évolutifs complexes".

introduced an intelligent reflection that increasingly mediated between flight and aggression.

"In individuals, the state of consciousness generated by associative memory surpasses that of unconsciousness, freeing them from their impulses (hypothalamic) and automatisms (limbic) through the play of their creative imagination, stimulated by knowledge and individual experience which motivate their reasoned actions, embedded in multiple rewarding collective interactions which have become necessary for their survival, reinforcing social cohesion".

This integrated system of the individual and society remains fragile, however, if the conscience does not manage to free itself from the mutilating conditioning, the most obscure ideologies and beliefs, which dangerously fix thoughts by inciting hatred and the worst destructive actions witnessed in History. This is all the more fragile if there is no understanding of biological revolution and the transformations of Civilisation as a complex process in which uncertainty predominates in the discrete relationship between the random variations that affect our species at random in the circumstances of natural and anthropic selective constraints that we produce through our mythical imagination rationalised into institutions.

Numerous popular beliefs and symbols are associated with the salamander, and these mystified descriptions possess a degree of truth that scientific knowledge is uncovering by explaining the illusions and fantasies that imagery the perception of these mysterious beings. In prehistoric times, a salamander carved on reindeer antler is mentioned at the Magdalenian (Upper Paleolithic) site of Laugerie-Basse, near Les Eyzies-de-Tayac in the Dordogne. This so-called command stick, adorned with a salamander, goes beyond the reality of an animal encountered in nature. Man stylised the head, body and limbs of the salamander, engraving the rugosites that he discerned on its skin; did he feel the heat of its burning poison on contact, to the point of totemising it?

Perhaps this is why this animal with aquatic tendencies has long been considered a familiar of fire and flames, the source of life and protectors, familiar to the Undines according to Paracelsus. For the Greeks, Triton was the son of the god of the oceans, Poseidon, and Amphitrite, who protected the sailors. Triton is depicted with a fish tail distinguished by a crete, and the Nereides lived with Poseidon in a palace near the lake Tritonis the aptly named. The guardian of fire for Christians, a salamander is clearly visible on a bas-relief in the central porch of Notre Dame de Paris, in the arms of a woman with hair that looks like long flames. Known as the vulcanal during the Renaissance, the salamander became the emblem of François 1° **"J'entretiens et j'eteins" ("I maintain and I extinguish").**

Subjected to superstition in the provinces of France, the deaf, because they had no apparent ears, were attributed the most diverse powers, including poisoning water and killing. The animal was sacrificed in cauldrons with mathematical filters by witches and often ended up raw in alchemists' stills.

We can see the value of exploring this long pathway of social instincts which, in the Euproctus, reveal the first beginnings of a potentially sociable relational arrangement during hunting cooperation. In the confines of a small pool, the unexpected reinforcement of the cohësion of Euproctes is initiated from each individual behaviour, from which emerges a collective strategy, the beginnings of what is considered to be the still primitive establishment of sociability.

This initial condition, during the course of a revolution, gave rise to a social organisation in other mammalian species, the primates and the human species in particular, which intelligently ensured the transfer of animal representations into rock art. These animals were hierarchised in frescoes painted or engraved on the walls of caves and on the totem poles of the first peoples, in a kind of elevation of thought that indicated the affirmation of the group's culture, a testimony to the emergence of the social bond resulting from increasingly sophisticated conscious cooperation.

A sign of power, the clan chief's salamander stick marks his ascendancy over the group of hunters, to maintain its necessary cohesion around a sacred symbol, in a way deified by shamanic mediation, indispensable for guaranteeing the social relationship, in order to confront nurturing nature by overcoming the ever latent instincts necessary for the survival of each individual and the clan. The link with nature has changed in the space of a few millennia, as a result of the interplay between biological revolution through natural selection and the singular emergence of this cultural edifice that structures the civilisation of distant origins, which manifested itself in prehistoric art using a few sharp pebbles, soberly arranged at Olduvai. Animal sociability underwent an original inversion, with a gathering of men, women and children, brought together by ritualised, transmissible associations, to become a social fact irreducible to natural selection alone.

In this way, the hunt is no longer left to chance, as in the case of the Euproctus, but is guided by the power attributed by the group to the bearer of the command stick carved with a 'salamander'. This animal symbol gives us an indication of the willingness of the men to cooperate under the formal direction of a clan leader who possesses the knowledge and power to lead the hunt and distribute the prey.

With this demonstration, we want to show the resolving power of the Darwinian reversal effect, this unique link between nature and culture, between biological revolution and the transformation of civilisation, is no longer inscribed in a clean break or an illusory continuity, but in the couple of a tendency moment of the social instincts which are reversed in the course of the history of animals by opposing natural selection, in the evolutionary succession of Primates, revealing civilised Humanity. This tendency to stemanciate from natural environments, while being dependent on climatic fluctuations and vital resources, is the progressive manifestation of a state of consciousness and an exceptional social fact that will be amplified with the development of the brain, the central nervous system, the upper and lower limbs...

There has been a liberation from the constraints of persistent primal impulses and the overly preoccupying tasks of survival that take over, by the reasoned conscience that has modërë the flight or tempered the aggression. Humanity has ennobled itself in the fertile imagination of artistic and cultural creation in all its forms, symbolically applied to the walls of caves.

However, the clan chief's dominance is not gratuitous; he must maintain his authority to guarantee the cohesion of the group around the 'salamander' totem. Instead of having a silver back like the dominant gorilla, his power is exercised by brandishing the symbolic stick and other attributes of institutionalised power. To maintain his authority, he must enforce the traditions, beliefs and laws that bind the group together. His hindsight guarantees him intellectual clarity, while the members of the clan immersed in piecemeal action find themselves distanced by their lack of knowledge of the hunting strategies he has integrated because of his hierarchical position. This power will be maintained until his death by the force of tradition and the initiation of the younger members, unless one of them, having acquired too much maturity and a competitive spirit, 'wishes' to take his place abruptly, finding in the natural selective model more satisfaction in cruel aggression than in soothing reflection and wise palavering around the fire.

Euprocts are still at the stage of competition for food, in a timid attempt to co-operate in hunting their prey. This stage includes a certain amount of animal sociability, which has enabled us to measure the amplitude of their social instincts and define a probability density for a tendency to the reverse effect, which we are comparing with that of mammals and hominin primates. This led us to write the final volume of the naturalist suite, **'Indri indri, voyage aux origines de l'Humanite'**, and the accompanying essay on the conditions of its disarticulation, which is seriously affecting civilisation.

We have reached the source of the Machiavellian plots, conflicts, crises and wars, of terrorism and terror, of the crimes that have unceasingly agitated human populations on earth from prehistory to the present day. During my operational sailings around the world, I was one of the players and a witness to these vicissitudes, just like our parents and grandparents who lived through the last world conflicts...

From biological evolution
to
the beginnings of civilisation.

From my first posting on the aviso-escorteur Commandant Bory in the Ocëan Indian Ocean, I keep in mëmoire the stopovers that helped to orientate my research into complex ëvolutive processes. We sailed from the Red Sea to the Mozambique Channel, from the Indian Ocean to the Persian Gulf, it was an initiatory navigation, a first periple that allowed me to perceive the biogëodiversitë.

As I became a citizen of the world, while remaining in the service of the flag flying at the stern of my aviso-escorteur, my gaze was still naively drawn to the desert spaces of the Horn of Africa. I was discovering the savannahs and forests of Africa and Madagascar, the still impërial Persian Gulf, and India, where the multitude of peoples and gods was revealed in the streets of Bombay, on the way to Ceylon... The animal life and civilisations of these regions left an indelible mark on me, and my subsequent cruises on the oceans, the Atlantic and the Pacific, where nuclear tests were carried out at the time, certainly reinforced my questions about the revolution of species and their extinction. But the preoccupation that occupied me more and more on a daily basis consisted of supervising and commanding sailors in the major conflicts that were beginning in the Mediterranean and the Middle East. During the reparative leave, the first surveys in the Pyrenees, in my early days as a naturalist, were the trigger for in-depth research, firstly into the theories of biological revolution. Aifeete on the aircraft carrier "*Clemenceau*", I would take an active part in safety and prevention on board the "*Tiger of the Seas*", engaged in non-stop overseas operations, alternating with embarkations on its counterpart "*Foch*".[45] ". My questions took on a whole new dimension during the Cold War, when the great powers were still fighting each other from the Red Sea to the Persian Gulf, Lebanon was exploding once again, Egypt and Israel were at war, Iran was changing regimes, South Africa was still under apartheid... Hence my question about the human race?

In the course of its ëvolution, Humanity has acquired a unique cultural island, in terms of the History of Peoples, civilisation is a fragile skiff, the very vulnerability of which I observed during my assignments on board the combat ships of the French Navy engaged in the oil war.

My meeting with Professor Patrick Tort enabled me to readjust and consolidate my reflections in order to work on a more exhaustive dimension of the theory of revolution by variation and natural selection, ënoncëed by Charles Darwin in conjunction with Alfred Russell Wallace. A scientific concept that I was to decline in in-depth research into the complex processes involved in biological evolution and the transformations of civilisation. In writing the naturalist sequel, I drew inspiration from Charles Darwin's journey to develop my method of analysis, forcing myself to reconstruct part of the long divergent trajectories of the animal species I studied and of the Hominid Primates in order to trace

---

45 On 1 October 1919, Marshal Ferdinand Foch, a native of Tarbes and Commander-in-Chief of the Allied Forces, arrived in Arreau to visit the house where his grandmother was born, and travelled up the Aure valley to the Saint-Lary hydroelectric power station. His name was given to two warships, a cruiser (1931) and an aircraft carrier (1960).

what Patrick Tort aptly called the reversive effect of revolution.

The beginnings of civilisation are buried in certain corners of the biological revolution of species, to a lesser extent and indirectly in plant species. On the other hand, instincts that tend towards a certain sociability are perceptible in animal species (corvids, baboons, chimpanzees).

By moving away from disastrous social Darwinism and reductive sociobiology, I embarked on fundamental research, necessarily applied research, such as that on fossil plants from the Carboniferous period.[46] The black ferns of the Cevennes hinterland are at the origin of the civilisation of coal, iron and steel; the Euproct of the Pyrenees and low biological energies; the Beaver of the Cevennes; the Lemurians of Madagascar. This naturalistic approach was to enable me to '*see*', with the help of models and simulations, how the divergent evolutionary transition of a species takes place and how the cultural singularity specific to the human species emerges, as it did in its civilisation, and is still emerging.

In the Hautes-Pyrenees, at the Moudang barns, our encounter with the Euproctes, **Calotriton** *asper asper*, challenged us. Their study required a considerable effort to understand the various ways in which they evolved, which raised many questions for us. The biological evolution of this Urodele led us to the frontiers between two worlds that were not unknown to us, that of water and that of land. This Lissamphibian lives in extreme conditions at altitude. **Calotriton** *asper asper* is an original species that breathes largely through its skin, and once it has developed, it undergoes metamorphosis and has the ability to regenerate severed organs.

70, 71 and 72. Euprocte des Pyrenees in its environment.

We reconstructed its phylogeny in order to provide a model for approaching its biological ëvolution, by observing it in its biotope. The aim was to understand its physiology and

---

46 This study looks at the socio-economic and ecological impact of coal and steel mining through the evolution of plants from this geological period.

metabolism, both aquatic and terrestrial, which are subject to strong natural selective pressures, to which are added anthropogenic constraints, which had to be identified. In the Moudang valley, on the banks of the Neste, the Euproctes live in small, shallow pools and hibernate on the ground, breeding in highly oxygenated streams. Their individual and collective behaviour is intriguing. To capture insects on the surface, they adopted hunting devices that I noted in my naturalist's notebook. By observing them closely, we were at the point where an animal's biological revolution was beginning to tip towards a primitive form of sociability, we were discovering the initial conditions of their instincts directed towards an organisation suggestive of a strategy, which could be interpreted as a nascent social organisation. How can we estimate this transition from individual behaviour to a sufficiently significant collective activity in an animal species, which will become an exclusive social fact in the Hominin primates?

The social instincts of Hominin Primates and Hominins in particular have been more assertive than those of other animal species, progressively opposing natural selection. The social instincts of Hominins have proved sufficiently effective to free them from natural environments. This notorious discrepancy accentuated their adaptability in a singular reversal from which Civilisation emerged, through an invaluable reversal effect.

The topological image of this process of biological evolution and the transformation of civilisation is the culmination of humanity's long evolutionary journey, represented not by an illusory, well-defined point, but by the area of the Mobius ribbon that develops until it twists.

If we follow this trajectory, we can see that in the course of its reversal there is no real break between omnipresent nature and growing culture, but a singular reversal that prefigures the advent of an irreducible social fact, in an irrevocable move away from the state of nature. According to this concept of the reversive effect of revolution, although civilisation has reduced natural selection, it has not cancelled it out; moreover, it has introduced a problematic accumulation of natural and anthropic selective constraints. Thus, the process of biological revolution associated with the transformations of civilisation produces as much, if not more, complexity, and this critical stochasticity has repercussions on the adaptability of species whose revolution has been compromised by our most harmful activities. Civilisation is flowing in this disturbed river of life, revealing waves of uncertainty at the origin of the crises that destabilise the governance of States. How can we assess these disturbances?

To simulate the tendency for the reversal effect, I have introduced a third reversal adaptativeness coefficient (*CAer*) into a multiplicative model that 'accelerates' the complex evolutionary process. This is a very powerful operator that characterises the area at which the civilisational reversal occurs, and whose positive or negative parameters, which we generate, interfere with natural selection. At present, these parameters, which depend on our socio-economic activities, have acquired the dangerous ability to interact simultaneously with biological variations and natural and anthropogenic selective constraints, introducing strong correlations between these three terms.

*CAbio (1-x)*CAdiv (1-x)*CAer (1-x)*

This function, pushed to the limits of stress and of the ecological valence of an animal species, makes it possible to define the amplitude of its instincts, which tend towards a form of sociality that is reversed by physical and behavioural compensations and overcompensations. By projection, we determine an amplitude of social instincts and a probability density of the reversal effect, in order to simulate it in three dimensions to estimate the tendency of the singular torsion corresponding to the organised collective behaviour of the animal compared with the exclusive reversal effect of the human species which serves as a reference unit. This coefficient in no way determines intelligence; it is designed as an indicator of our persistent and excessive relations with the natural environment, so that we can better refocus our most deleterious activities on low biological energies, to maintain the vital coherence of species by promoting the essential social cohesion that sustains civilisation.

Isolated for thousands of years, the Euproctes have remained confined to each valley without undergoing any major changes. Our observations in the Hautes-Pyrenees and the lengthy research they prompted have given us access to their phase transition between the aquatic and terrestrial environments.

At the end of this exploration of the Moudang valley, we assessed the probability of evolution of the Euproctes, for which we used the biological adaptability coefficient *CAbio = 4.39* of the Anurans and Urodeles and the divergence coefficient of the Urodeles between *CAdiv = 0.32 and 2.74, which* corresponds to a relatively stable evolution.

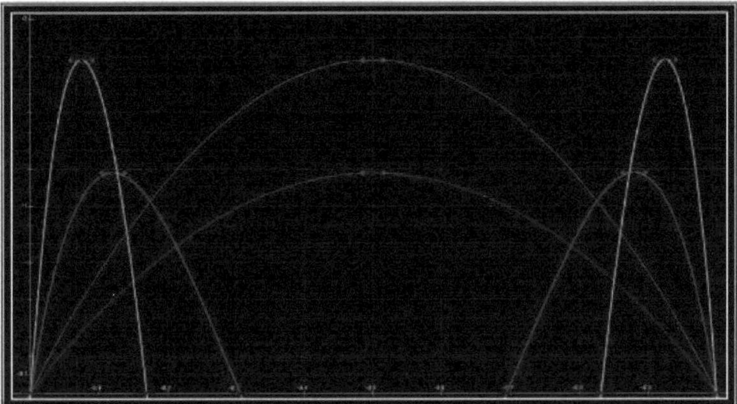

73. Simplified approach, divergence of Anurans and Urodeles.

In the absence of sufficient records to simulate their probabil^ devolution, the test modèle is weightedёrё by binary corrective factors (0 or 1) which are introduced into the following additive function.

CA bio (1-x) + ((CA div=(A+B+C+D+E)/5))(1-x))

In the light of the preceding paragraphs, it is possible to act on the traits that could lead to this evolution, for example by favouring to the extreme either terrestrial life or the more aquatic life that is currently dominant for the Pyrenean Euproct species, for a *CA bio = 1.12* corresponding to an asymptotic stabilising evolution. To simulate the biological devolution probability of the Euproctus we forge the modèle :

-    Cutaneous respiration would be less terrestrial *A= 0*, more aquatic *A=1*.
-    After reproduction, during development, we consider that the segmentation process from egg to larval stage with gills is more spёcifically aquatic *B=1* than terrestrial *B=0*.
-    Mёtamorphosis tends to be more terrestrial *C= 1*, than aquatic *C=0*, although for the Euproctus this is debatable.

- the ге́дё^^^, would be less terrestrial **D=0**, than aquatic **D= 1**, if we take into account its greater efficac^ in the larva than in the adult, the aqueous context seems essential.
- Behaviour, less terrestrial **E=0**, more aquatic **E=1**.

Parameters/trend	Land	Aquatic
A. Skin respiration	0	1
B. Development	0	1
C. Metamorphosis	1	0
D. Regeneration	0	1
E.Collective behaviour	0	1
(A+B+C+D+E)/5	1/5	4/5
Organic sales div	0,2	0,8

74. Summary table of the parameters[47]
of the coefficient of adaptability of Euproctes.

For the terrestrial solution: *1.12x (1-x)+ 0.2 x (1-x).*
And the more aquatic solution: *1.12x (1-x)+ 0.8x (1-x).*

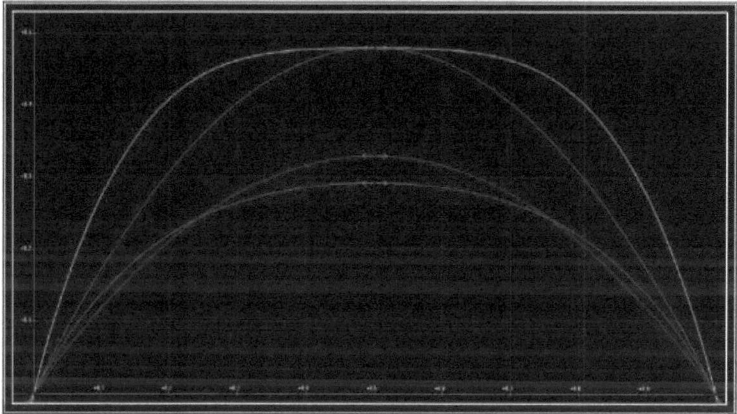

75. The terrestrial solution does not diverge (red and magenta), but the aquatic solution (blue and green) shows a trend that is confirmed by a slight divergence that is still unclear, the first signs of possible speciation.

The summary simulation shows an evolutionary trend that seems to be more directed towards the aquatic environment than the terrestrial environment, which would confirm a fairly stable biological evolution with the possibility of spëciation that would have to be validated by assimilating real data to make the model more reliable. However, the impact of minute random variations subject to fluctuations in cumulative natural and anthropogenic selective constraints are still likely to act on these two devolution modalities and influence its divergence. On the other hand, too rapid a warming of the climate will radically wipe out the species, unless it finds transitory refuges on the seabed at high altitudes and in protective cavities at lower altitudes. This is what has probably happened in the past, when geological, climatic and ecological fluctuations helped to isolate populations of Euproctes in watercourses that have ensured their survival to the

---

47 This type of matrix is useful for working on the correlation factors of genetic and morphological variations in relation to natural and anthropogenic selective differential constraints.

present day, without major modifications.

As a rough guide, the trend in the reversal effect of this species would be :

*4,39x(1-x)*1,12x(1-x)*0,32x(1-x)*

This gives an amplitude of social instincts of *0.47* and a probability density of tendency to the reversive effect of *0.0012*. A very low value compared with the reference unit.

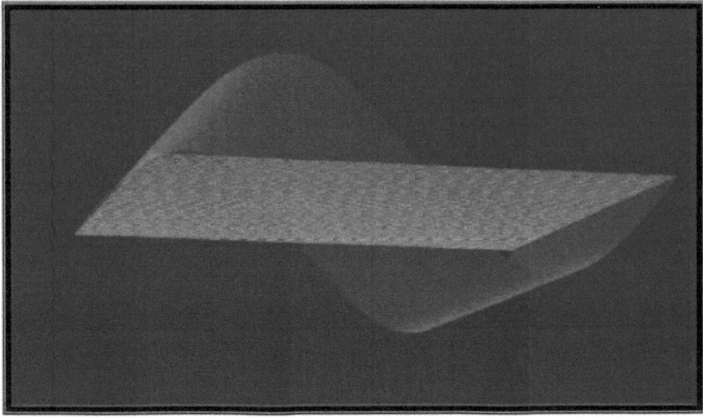

76. Simulation of the trend towards the reversal effect of Euproctes (blue), very slightly downgraded compared with the civilisation of the human species, the reference unit (red).

In our models, we have established a link between the past and the present, starting with the phylogeny of the Lissamphibians and Salamandrids and ending with the Euproctes. Using the adaptativeness coefficients, we have an exhaustive model for approaching its evolution. To obtain a test model, we corrected these adaptativeness coefficients, which summarise the initial conditions of the biological revolution of the Euproctes, by weighting them using corrective factors from our study of skin variations at altitude, egg segmentation, metamorphosis, regeneration and social behaviour in the water. For a very low level of confidence, we give a probability of evolution for **Calotriton** *asper asper* in the Pyrenees. This species should continue to differentiate locally, with the possibility of more assertive speciation towards the aquatic environment.

To conclude, we formulate a hypothesis on the capacity of the Euprocte a гёдёпёгег an organ sectionnc' accidentally or by a predator. Considering the ëtages of its phylogënie and its predominantly aquatic probabilik devolution, considering its ontogeny and mëtamorphosis, which include the devices prerequisite to its гедё^^^. From these initial conditions dëmontrëes in our models by coefficients of adaptativite more orientës towards the aqueous medium, where the гедё^^ has optimisëe at the level of cuta^e cells: The latter possess respiratory and excretion autonomy indëpendent of the дё^гак circulation via the intermëdiary lymphatic system, with a major role in nuckation and cell formation. This localisë process would be under the dual control of the cells' genes in co-action with the prokine complexes of the mitochondria, as well as the nervous system, which would cause the contractions of the lymphatic creurs, locally reactive to the frequency of the proton and photon field, pu^ a threshold of the mitochondria. In the event of an incidental amputation, shock and the immune system would modify the amplitude of this local field, unik of coordination of the disrupted cells, by soliciting the stem cells by activation of their ë^^κ minimum.

This activating and modërateur process at low biological ë^^κ would induce their divergence and cell division, this aggregate formation being constituted in blaskme, to initiate the rëgënëration of the organ. These so-called morphogënëtic interactions can be likened to the difkrential couplings of threshold pulse fields, intra- and inter-cellular, as coordination uniks at the quantum level, which are deployed during chain reactions, in a mosaic of structuring biochemical and hydro-thermo-dynamic interactions, rëgënëranting organs that become functional again.

By moving away from the aqueous environment, the amplitudes of the cellular coordination fields changed, causing a loss of autonomy of the cuUinc'es cells in favour of the gënëral circulation for respiration and excretion. This complex evolutionary process reduced the capacity for rëgënëration of the more terrestrial species, which became practically non-existent in mammals.

As an experiment, we simulated the reversal effect by defining a coefficient of adaptability complementary to the biological one and of divergence in order to measure the amplitude of instincts at their stress limits, in order to detect a probabilik of the onset of the beginnings of a tendency towards the reversal effect. The behaviour of Euproctes points to the beginnings of a sociability induced by prey capture practices, individual acts of hunting that appear to be organised in collective battues.

These are the first manifestations of a very distant pseudo-culture which, after a long trend reversal, will indeed lead to a reversal in the Hominin Primates (Tort, 1983), which remains decisive for the Human species, whose origins we will have to probe in the last volume of the naturalist suite devoted to the Lëmurians of Madagascar in the company of *Indri indri*, in a journey to the origins of the Humarnk.

Before I finished writing the manuscript for this second volume of the naturalist suite, I learned that half of the 2021 Nobel Prize for Physics had been awarded to the Italian Giorgio Parisi:

"For the discovery of the interaction of disorder and fluctuations in physical systems, from the atomic to the planetary scale".

This prize for excellence in complex processes reinforces my exploratory approach, a naturalistic investigation far removed from the publications of this brilliant researcher and others whose theses I have consulted at length, too little known to the public, and which are mentioned in the text and bibliography.

In writing "L'Euprocte des Pyrenees" and the accompanying essay "Les basses energies biologiques", I demonstrated the stochastic adaptability of a species whose fundamentally random variations are confronted with the randomness of natural and anthropic selective constraints.

In terms of scholarly publications, these promising bitters will shed light on my navigation as I write the last two volumes of the naturalist suite and my latest synthesis on the theory of contingent revolution. A long-term exploration to probe several aspects of the divergent transition between linearity and non-linearity in order to justify my definition of adaptability, and ultimately validate this operator as the keystone of complex processes in biological evolution and transformations of civilisation.

Appendix 1

A reminder of complex evolutionary processes.

In ëvolutive processes described as complex, because of the many interactions that take place at different levels of integration of living organisms in the organisms of the individuals that make up the population of a species subject to natural selection by its environment, by definition.

"The Darwinian relationship of adaptability, from its fundamentally random variations to the randomness of selective constraints, is stochastic in nature".

If we consider that, at the origins of a species, the initial point of accumulation of this ensemble unfolds and is modified during its phylogeny, its contingent evolution is likely to produce divergences. The evolutionary processes of species are reconstructed using fossil forms and their descendants up to present-day living species. Cladistics is a method of scientific classification of species designed to build exhaustive trees whose resulting phylogenetic ultra-metricity is characterised by singular degrees of stochasticity. The approach modelling of these tree diagrams enables us to estimate the transition between linear and non-linear during the divergent trajectory of a species using a standardised biological adaptability (*CAbio*) and divergence (*CAdiv*) coefficient. On the other hand, we have shown that random effects induce discrete underlying consequences, which are highly structuring and/or de-structuring, in linear modes with damped fluctuations and during phases of intermittency, which occur during fully developed non-linearity in chaotic form. The relative stability of living organisms, which can be observed at different levels of living integration, conditions the life and evolution of species if, and only if, the components (waves and particles, atoms, molecules, proteins, cells, animals) of these biological structures acquire, by virtue of their physical and chemical properties, which organise their biochemical affinities: **A "vital coherence"** which, in our opinion, would be initiated from a genetic and epigenetic mosaic maintained by a low biological energy, by a coupling of waves, considered as the cellular coordination unit, produced at the level of the mitochondria, by an intermembrane proton pulse field that has reached a saturation threshold of protons and photons.

On its evolutionary trajectory, a species that can be represented by a coefficient of biological adaptability (*CA bio*) and divergence (*CA div*) is statistically out of stable equilibrium in a biological state of homeostasis around the average of its ecological and biological valence that maintains its vital coherence, at the limit of aggression (stress) and letality. When it reaches a state of unstable out-of-equilibrium, beyond the zones of compensable and over-compensable stress at the limits of its viability, by breaking the linearity of its trajectory, during internal random variations in their hazardous relationships with external constraints, divergences appear favouring its diversity or its extinction by natural selection, highly anthropised selective constraints.

Thus, a species develops, reproduces and ëvolves by differentiating itself in space and time, while its cellular coordination unit, the threshold pulse field, would manifest itself in short- and long-range space-time via its divergent derivatives. Its evolution, summarised very synthetically in a tree diagram, can be quantified at each bifurcation by re-evaluating its degree of stochasticity using adaptability coefficients.

The universal nature of this normalized transition property, between linear and non-linear, means that we can model and simulate the biological evolution of a species at several levels of resolution.

- Over the long term, using an approach based on a pre-established ultra-metric tree structure, corrected by the phylogeny of a species, this operation provides access to coefficients of biological adaptability and divergence representative of the revolution of a species from its origins to the present day.

- Short-range in a test model, assimilating updated data for the species under study, to correct the coefficients of biological adaptability and divergence of the approach model.

Using the test model, it is possible to explore the evolutionary potential of a species in order to study its most likely trajectories, and to experimentally push the simulation to the limits of its evolutionary trajectories.

In this way, the biological adaptability coefficient of the approach model derived from the phylogeny of a species is completed by the divergence coefficient corrected by updated factors. These coefficients of biological adaptability and divergence are introduced into an operational test model, with which successive iterations and reiterations are used to obtain divergence sections that indicate the probability of its evolutionary trend. The choice of parameters for species variation and natural and anthropogenic selective constraints is decisive in obtaining adaptativeness coefficients of acceptable reliability in relation to a confidence level set during modelling and prospective simulations.

More specifically, for a species under study, the complex evolutionary processes are formalised in logistic equations in order to obtain a probability of their evolution by divergence. Using the biological adaptativeness coefficients (*CA bio*) from the approach model combined with the calculated divergence adaptativeness coefficients (*CA div*), we assimilate the data by introducing them into a sum of logistic functions for a population X representative of the species under study.

CA bio X (1-X) + CA div X (1-X)

This function, iterated by successive transforms, gives a cross-section of biological revolution with its divergences representing probable trajectories that can be carefully extrapolated by pushing and justifying the simulation parameters to estimate the evolutionary potential of the species under consideration.

As we do indeed know the critical thresholds for the values of the adaptability coefficients that cause bifurcations, this simulation is interesting in that it is possible to act either on the variations in the species (genetic, morphological, behavioural) or on the selective constraints (natural, anthropogenic), or on both, by making judicious use of the correlation factors.

However, this requires precautions and an understanding of the biological and ëcological processes of the species in question, bearing in mind this warning.

"For low values of linearity (less than two), particular importance should be attached to damped fluctuations, which nonetheless induce underlying discrete structures".

We find this problem in the intermittent phases of full chaotic development, apparently linear, but with highly variable underlying structures, whose fluctuations and disturbances must be understood at the level of vital coherence in cases of degenerescence and pathological anarchic development. The result of the test model is used in the form of a probability of evolution within a confidence interval, and is based on the causes suspected during the observations, which are used and verified in the light of current scientific knowledge. This exploratory method requires multidisciplinary skills and research resources that go beyond those we have used to solve the case of complex evolutionary processes such as that of Euproctes.

To conclude this reminder, we can use a model approach to initiate an exhaustive diagram of the revolution of a species, from its origins to the present day, in order to obtain coefficients of biological adaptability and divergence. Then, using the coefficients obtained in this way and those derived from observations, after correction, produce a test model to simulate and project the probability of evolution of this species. In addition to this stochastic aspect, the highly random nature of non-linearity reveals underlying bounded structures that are difficult to predict. Nonetheless, they are decisive in maintaining the vital coherences that condition the relative homeostasis of the organism of the individuals in the population of a species. At the limits of linearity and non-linearity, understanding the nature of these bounded structures appears to us to be of prime importance in the treatment of the theory of revolution extended to that of the transformations of civilisation by means of complex evolutionary processes.

This has led us to complete this modelling of biological revolution, by extending it to its anthropological concept, by simulating the reversal effect (Tort, 1983) to trace the origins of the social instincts which, in certain animals, manifest themselves in a proto-culture. This result of the unfolding of consciousness and cultural emancipation, specific to the human species, has manifested itself exclusively in a trend reversal that clearly distinguishes it from other animal species, through the expression of Civilisation. In the reversal effect model, we use three adaptativeness coefficients, biological (*CA bio*), divergence (*CA div*) and reversal effect (*CA er*), introduced in three multiplicative logistic functions. By means of its first and second derivatives, we define the maximum amplitudes of social instincts at their limits, as well as the probability density of the reversal effect that results from this, in relation to the normal curve of their biological evolution by natural selection, which reverses to produce civilisation, exclusive to the human species, the beginnings of which we have explored in the animals studied successively, the amphibians, rodents and primates.

Although the universal character of non-linearity enables us to resolve complex evolutionary processes, each research project has its own specificity, and requires a different approach for each case of study, with the usual precautions, whether it be the use of phylogenetic trees or, more particularly, the choice of parameters that enable the coefficients of adaptativeness to be updated by assimilating observational data. In the essay on low biological energies, I explained the difference between chance arising from selective circumstances, which differs markedly from the random aspect of variations, due to the intrinsic properties of inert matter, and by extension to the combinatorial nature of living beings.

In the case of complex evolutionary processes, such as embryogenesis, knowledge of the initial conditions is essential to adjust the test model, which is highly dissymmetrical in the first segmentation phases and highly correlated in the subsequent phases for the formation of organs, which are formed in the course of a chain reaction in a mosaic of interactions simultaneously contained by gene co-action and the energy produced by protein complexes.

The choice of parameters to be assimilated is decisive, both for variations and for selective constraints, so it is not the quantity of parameters to be assimilated that is important, but understanding their nature and the interactions that are likely to produce non-linearity. Using the Gaussian distribution of the population of a species, we can break down the divergence process: when a disturbance modifies its shape, for incidental variations in the small values, the normal curve develops exponentially; this causes modifications to its mean and its out-of-equilibrium standard deviations, the instability of which increases towards the large values until it causes a significant decoupling of the divergence of the species under consideration.

Appendix 2

Phylogeny of Amphibians and Urodeles

As *Vertebrata, Sarcopterygians, Rhipidistians, Tetrapods and Amniotes, Lissamphibians* have remained uninhabited by the aquatic environment, so that present-day forms differ little from ancient forms and possess morphological and physiological characteristics that give us indications of the transition between the aquatic and terrestrial environments during their evolution. This transition was expressed by changes, particularly in the mode of reproduction and development, during a necessary metamorphosis in which gill respiration was replaced by mouth-pharyngeal respiration and lungs, sometimes non-existent or rudimentary as in the Euproctus, which maintained cutaneous respiration. These systems are supposed to ensure respiratory continuity between two environments, but it is mainly cutaneous respiration, supplemented by oral-pharyngeal respiration, that performs this transitional function. It is therefore of particular interest to trace the phylogeny of the Lissamphibians in order to understand the main morphological and physiological variations that have affected the class of Batrachians, Anurans and the order Urodeles in particular.

From the primary and early secondary periods onwards, the *Temnospondyla are* closer to the *Anurans*, and this group includes the *Batrachomorpha* and the *Antracosaurs, which are* closer to the reptiles, while the *Urodeles,* which date from the secondary period, are classified separately.

Among the *Batrachomorphs,* we distinguish the genus *Ichthyostega* from the Upper Devonian (Greenland, 360 Ma), which gives us an idea of the morphology of the first amphibians. It measured between 1.20 and 1.50 metres in length and differed from the *Rhipidistians*. It looked like a heavy salamander that had retained bony plates on its skull and a covering of scales on its belly. It had a flipper and six fingers on its front legs.

Among the *Rachitomes* of the Lower Carboniferous, which extended into the Triassic, there were aquatic forms and some that were more terrestrial, equipped with solid dermal armour; the hand had four fingers and the foot was reduced to five. In *Dinosaurus,* the gill skeleton is preserved in the adult stage, and this neoteny is observed in the larvae of today's batrachians.

Following on from the Rachitomes, in the Lower Triassic, came the *Trematosaurs, which lived* exclusively in the sea. Derived from the Rachitomes were the *Stereospondyla, which were* very aquatic, with limbs that were not very ossified for terrestrial movement, and equipped with gills and webbed front legs.

The anurans, frogs and toads have been known since the Upper Triassic period, when *Triadobatrachus* was discovered in Madagascar. It was barely ten centimetres long, had a small tail and hardly jumped at all, while others were distinguished by a marked aptitude for jumping and showed adaptations to a wide range of environments, including extreme drought and cold.

Although the larval life of ***Lissamphibians*** remained aquatic, after adopting different degrees of mëtamorphosis depending on the species, the adult tended to continue its activities near water, remaining in very humid environments. Let's take a look at the ***Embolomeres*** and ***Seymouriamorphes*** groups, which derive from the ***Osteolepiformes***. The ***Embolomeres*** are amphibians from the Lower Carboniferous: the predominantly aquatic ***Pholidogaster*** had a long body extended by a powerful tail and very short limbs. ***Seymouriamorphs*** are Upper Carboniferous and Permian forms that are thought to have originated from ***Embolomeres***. They are less aquatic than the Embolomeres, ***Diadectes*** and ***Seymouria*** had very extended legs under their massive bodies extended by a tail, they would have acquired a probable adaptation to terrestrial life by a herbivorous diet of aquatic plants and were probably endowed with aptitudes to move on the ground in search of terrestrial plants. Fossils of ***Lissamphibians*** from the Devonian and Carboniferous periods show significant changes in their morphology:

- The bones in their legs thicken, and powerful muscles enable them to move without dragging their belly as they move from water to land.
- The spinal column becomes more robust.
- The eardrums change and pick up sounds in the air.

350 million years ago, ***Lissamphibians*** acquired terrestrial aptitudes while remaining dependent on the aquatic environment from which they tended to emerge from the Devonian onwards (fish with lobed fins), dividing into two groups in the Carboniferous, the ***Labyrinthodonts*** and the ***Lepospondyls,*** probably the ancestors of today's Lissamphibians, which from the Permian to the Triassic gave rise to three groups:

- ***Anurans***.
- The ***Urodeles,*** which includes nine families of newts and salamanders.
- ***Gymnophiones***.

As for the ***Urodelomorphs***, whose fossils differ very little from present-day forms, their classification is still the subject of debate among paleontologists and systematists, who give us their main characteristics:

- The skull has undergone numerous bone reductions.
- However, the endocranium remains cartilaginous.
- Many bones are missing (suborbital, parietal).
- The upper jaw is well sutured to the skull.
- The gill skeleton is always developed.
- The limbs are small.
- The skeleton has an imposing vertebral column of around one hundred vertebrae.

In the revolution approach model for ***Sarcopterygians*** and ***Rhipidistians,*** we provide the most exhaustive representation for assessing the coefficients of adaptability of Lisamphibians and Batrachians, Anurans and Urodeles in particular. In the ***Caudata*** order, the ***Salamandridae*** family, according to the current classification, comprises around twenty species.

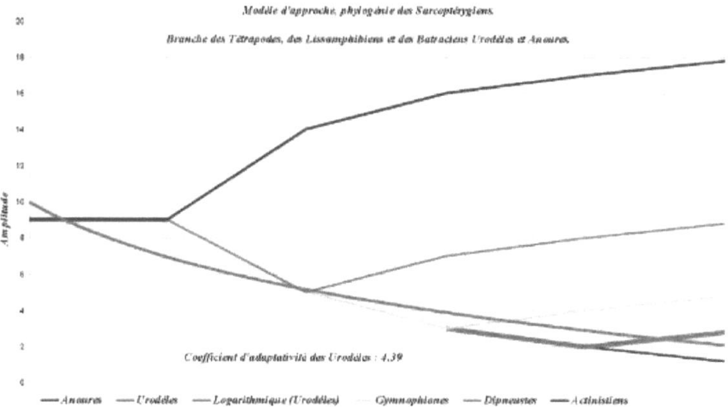

77. On the tree diagram, the top branch is clearly detached and is that of the Actinistians (in dark red), below which the Rhipidistians branch off and diverge into the Dipneustes (green) and the Tetrapods (yellow). The divergence of the Tetrapods is divided into two paths, that of the Lissamphibians (yellow), and the unrepresented Amniotes, which divides between the Mammiferes and the Sauropsids. The Lissamphibians branch includes the Gymnophiones (yellow) at the top and the Anurans (blue) and Urodeles (thick red) at the bottom.

For the Urodeles branch, the long-range biological adaptability coefficient is estimated at 4.39 (red). Still at long range, the coefficients of divergence adaptability of the Urodeles branch would range from 0.32 for a stabilising evolution without major modifications to 2.74 if we consider their exponential evolution, which would tend to diversify. In the absence of observational data, we have carried out prospective simulations by forming the variable characteristics of the trends in its respiratory abilities, which are more aquatic than terrestrial.

Lissamphibians are thought to date from the Lower Triassic (240 Ma), and to have acquired cutaneous respiration, which was absent in the first scale-covered tetrapods, seriously limiting transcutaneous gas exchange. The **Crytobranchoidea** suborder of primitive salamanders is currently represented by the Chinese salamander (**Andrias** *davidianus*) and the Menopoma (**Cryptobranchus** *alleganiensis*) from North Americaq ;

As adults, *these* species have larval characteristics such as.

- The gill slits.
- eyes without eyelids.
- They breathe thanks to the many folds of skin that are very thin and highly irrigated by capillaries.

The suborder **Salamandro'idea** is made up of **Plethodontidae**, without lungs or gills that breathe through the skin, while the essentially aquatic suborder **Sirenoiddea** resembles an eel without hind limbs, while the forelimbs are reduced. With the exception of a few frogs that live in the brackish waters of mangroves, Lissamphibians do not live in the dirty waters of seas and oceans.

The main common characteristics of Lissamphibians :

- They are poi'kilothermic, meaning they are unable to regulate their temperature by adapting to the climate and weather conditions, which affect their low metabolism and behaviour. Their reproduction, growth and metamorphosis are sensitive to temperature variations via the pituitary and thyroid glands. Their survival in the cold involves hibernation at very low biological energy.

- Their water-permeable skin facilitates gas exchange, enabling them to remain in the water. Terrestrial abilities vary from species to species, with the production of a layer of mucus and the acquisition of additional respiratory adaptations.

- Serous glands produce toxins, the parotid glands produce a neurotoxin such as bufo toxin, and the salamander has this type of gland behind its eyes.

- Skin colour depends on chromatophores implanted in three layers of cells.

- The respiratory system changes during metamorphosis. In the larva, the two-chambered gullet propels the blood towards the gills and the rest of the body. After metamorphosis, in the adult, the ventricle has one ventricle and two auricles, and it is the ventricle that sends the blood to the lungs and throughout the body in addition to the lymphatic system.

- The nervous system resembles that of fish, with the cerebellum controlling heart rate and breathing, the mesencephalon coordinating muscles and the telencephalon picking up olfactory signals, sight and vibrations. Since hearing is not very effective, the telencephalon plays a role in learning.

The ossified skeleton of Urodeles consists of a pelvic girdle and a scapular girdle on which the four short locomotor limbs are implanted. The head is broad and the teeth pedicellate, and the trunk is extended by a tail. Aquatic larvae have gills and undergo metamorphosis.

Appendix 3
Notes and chronology of observations.

Euproctes were first observed in July and August 1983, and again in 2006 and 2009. The water basin in which we observed the Euproctes in July and August was located above the Moudang barns at an altitude of 1,600 metres. The Neste flows between the Pic de la Hount and the Pic de Pene Abeilliere, following the path that runs alongside it towards the Pic de Bataillence on the Spanish border. The observation zone is located from the ferruginous spring that feeds the Neste and the Moudang barns in a rocky valley that is marshy in places. The temperature varied very quickly depending on the summer weather conditions, from very hot to cold when it rained or when it was foggy, with humidity ranging from 20 to 100%. In fine weather, temperatures averaged 17°C in the morning at around 10:00, 23°C at 12:00 and 24°C in the shade at around 14:00, with over 35°C in the sun.

With an outside temperature of 14.5°C, the water flowing into the pool at Euproctes was between 10°C and 13°C, varying according to the depth of the pool (between 10 and 25 cm), the flow rate, and the nature of the bottom and beaches, made up of rock, pebbles or gravel, bordered by vegetation. At the bottom, under a small rock, the temperature is 11°C in the shade. Knowing that mating takes place at 12°C with a search for highly oxygenated water. Normally basic on the Neste, the pH of the water in the pond was 6.8 to 7.5 for a resistance of 0.2 kiloohms and 59.8 micro-amperes. The pools were very ferruginous at the source, with the iron content decreasing with distance from the source.

Recorded on 3/7/1983 :

General observations on the road between the Moudang bridge and the Moudang barns on the way up to the source of the Neste du Moudang.
- Two ducks.
- Euproctes.
- Red frogs.

Recorded on 2/8/1983 :

General approach observations on the same route.

At an altitude of 2,500 metres, the weather was very fine and windy in the evening. The air temperature was 17°C at 9.00 am and 22°C at 12.00 pm.
- Four griffon vultures in flight over the ridges
- A yellow-billed duck
- Euprocts and frogs.
- A viper crosses the road.

Recorded on 4/8/1983 :

General observations on the same route.

Overcast with ëclouds, thundery evening, overcast very cloudy and low ceiling. Air temperature between 14°C and 17°C:
- Two vultures in flight over rocky outcrops

Recorded on 7/8/1983 :

Above the Moudang barns, going up to the source of the Neste du Moudang alimentëe by a waterfall, in a small basin of calm water, with a rocky bottom covered in mud and dëp6ts we observe Euproctes. The overhanging waterfall, which feeds it accentuates the oxygënation of the 10 cm to 30 cm deep pool. The bank is lined with rocks, gravel, moss and plants, and the gentle current discharges the water into the torrent. The average air temperature was 11.5°C, and the average water temperature was 10.5°C.
- Euproctes move through the water by hiding under rocks.
- An Euproctus oxygenates itself in one of the small reservoirs of the fast-flowing stream that feeds the Neste, with its upper body and head out of the water.
- In the hollow of an underwater rock, a male surrounds a female under her head, and they stand motionless.

Recorded on 9/8/1983 :

Basin of the ferruginous spring on the Neste du Moudang. Fine weather with wind, rain and sunny spells at the end of the evening, temperature between 17°C and 22°C.

12h00: Above the peak of Garlitz (2798 metres), two griffon vultures fly over a wooded and rocky area.

13h00: The group grows to eight, then ten griffon vultures. They fly over the valleys in circular flights at varying altitudes, heading towards Les Cretes.

16h00: Two vultures in flight over a rocky area.

16h00: two common kestrels in flight over the Pic de la Hount, hovering and heading towards the nest, they land on a rocky ledge above a wooded and grazing area. One of the kestrels captures a prey item on the ground and flies off, the other joins him and seizes the prey in flight and heads for the nest.

Recorded on 11/8/1983 :

Pont de Moudang, route des granges. The cloud ceiling was low, then the sun and wind rose around 12.00.
- Eight Griffon Vultures in flight on the ridges.

- Ten groups of isards on the pastures near the road and two at altitude.
- In the basin of the ferruginous spring: Two Euproctes intertwined at body level under the front legs. The female is immobile, belly up, the coral colouration of the belly can be seen, the periods of immobile resting last up to three quarters of an hour, before starting a movement conducive to fertilisation. After several attempts, the male becomes agitated and after a brief contact separates from the female. The female leaves the middle of the pool, followed by other males, heading for the entrance to the more flowing water. The mating area is located in the middle of the less agile pool or squarely in the stream sheltered by large stones.

Recorded on 12/8/1983 :
On the Neste, above Les Granges (1200 metres), very low ceiling, light wind.

Record of 13/8/1983 :
Air temperature at 2pm: 24°C in the shade, 32.5°C in the sun, light wind.
Progress towards the Aumar pass (2400 metres), exploring the scree under the neves, in full sunshine, observing the high altitude grazing areas (cattle, sheep, horses).
- Two choughs on rocks opposite the Neouvielle, then on scree.
- Two vultures in flight over the Neouvielle.
- In a lake, many tetards in very clear water.

Recorded on 16/8/1983 :
On the Pont de Moudang road, between 1000 and 1600 metres, the average temperature was 17°C, with no wind, thunderstorms with sunny spells and a low, overcast ceiling at the end of the evening.
- Crows, jays and griffon vultures.
- At 12:00 noon, four griffon vultures in flight above the cretes at altitude. One adult and three smaller ones searching for speed above the summits. One small one joined the larger one, they intertwined their talons in flight and circled together, letting each other fall, then separated. They resume their ascent flight and repeat this scene twice.
Other circuits will be made towards the cirque of Troumousse and on the Neouvielle more frequently.

Data as at 1/8/2006 :
15h00, fine weather, fog at altitude in the afternoon.
- Above the barns, several vultures circling the peaks.
On a tributary of the Neste, in a waterhole, four Euproctes (around 10 cm) were hunting green grasshoppers. The yellow dorsal line is visible, the belly is medium red to coral.
- At 2,400 metres, the temperatures in the water hole were as follows: Running water on entering the basin, 10°5 C. At the bottom of the water, under rock, 11°C. In the water in the shade, from the bottom to the surface, 15°C to 24°C. In the sun, at 12:00/35°C, 15:00/20°C, 17:00/11°C. Pool water PH 7.4.
- They are more attracted to live prey, lighter rather than darker, and are more reluctant to capture other insects.
- The hunting tactic consists of moving along the rocky bottom in the direction of the prey that is stirring on the surface, then propelling itself upwards in its direction, waiting and floating. When the prey passes overhead, it catches it in its jaws and sinks to swallow it at the bottom. During this withdrawal, another Euproctus tries to swallow its prey, the attack continues with an embrace similar to mating using the tail and they bite each other. Then they move off, one of them catching the sinking insect.

Recorded on 2/8/2006:

Cloudy and sunny.

- 13H00, at the water hole, an adult and a smaller one dëplacent at the bottom.

- At 3pm, the adult captures a grasshopper on the surface, eats a leg and swallows its prey whole.

- The little Euproctus with the yellow dorsal line tries to get out of the water and falls back. It manages to dëplacer towards the bank, gets out of the water, crosses a flat stone for 50 cm and takes refuge under a clump of vëgëtation.

Recorded on 7/8/2006:

Very fine weather, light wind.

- At 2.30 pm, an Euproctus with no apparent yellow line comes to gobble up an insect on the surface.

- Four Euproctes are hunting, one of them floats by inflating its belly, after three gouges of air at the surface, it waits for its prey between two waters, when it passes it captures it at the surface or after several unsuccessful attempts, tired it drops to the bottom releasing air bubbles and waits for another insect.

Recorded on 8/9/2009:

Observations centred on the streams and basins at Euproctes, on the Neste du Moudang and opposite the source of the Reine towards Port de Moudang.

- Two Euproctes entwined in the current.

- A larva on the moss.

- Six Euproctes hunt insects that fall into the pond.

- Some forty vultures above the Pic de la Hount

- Source ferrugineuse de la Reine, no Euprocte but very well-developed aquatic plants.

Recorded on 9/9/2009 :

Very fine weather, ferruginous spring on the Neste du Moudang, presence of Euproctes with gills covered in ferruginous particles.

Appendix 4

Impact of altitude

on

cutaneous respiration in amphibians.

At a constant temperature (12°C), if we consider the liquid external environment, the epidermis is simultaneously in a liquid state at the level of the water film on the surface, becoming more crystalline in the first layers of cells. In the dermis, the connective tissue appears as a semi-crystalline amalgam. These three biological phases are in an unstable equilibrium around the rëfërence tempërature conducive to the life and reproduction of Euproctes.

Depending on the change in altitude, the cuUine tissue changes, and the environment acts on its phase space, as J.A. Vellard found out when he ëШ^ë the adaptation of batrachians to high altitudes in the Andes, between three and five thousand mëtres, which he compared to low-altitude species. Up to two thousand metres, altitude has little effect, but in this region with its temperate and cold tropical climate, above 2,000 metres, the dryness of the air, the drop in atmospheric pressure and the fall in oxygen tension, as well as the temperature, contribute to temporary adaptations, if not definitive adaptation. The effects of radiation on the photosensitivity of the skin and its pigments, which capture light by reacting to the intensity of ultraviolet radiation, should also be considered. On amphibians, the naturalist J.A. Vellard notedë at high altitude :

\- In Andean climatic conditions, a thickening of the skin and the formation of corseted cuticles, pustules are very dëveloppëes and the cornification of the epidermis is very marked.

\- A gradual adaptation to life in humid and semi-aquatic environments, becoming systematically aquatic at very high altitudes.

He concluded that altitude at these latitudes has morphological and physiological effects and modifies the biological cycle.

\- At 2,300 mëtres, the skin is dry and warty in a humid environment.

\- At less than 3,000 mëtres, the size is great, the life partially terrestrial, the skin ëpaisse et coriK'e, the choanes are reduced, the tympanum well Аоттаё.

\- Between 3,000 and 4,000 mëtres, the tendency becomes more aquatic than terrestrial, the skin is less thick and becomes granular depending on the tempërature. The choanae are more than reduced and the eardrum is incomplete.

\- Finally, at 4,000 metres, the adaptations are remarkable, life is entirely aquatic, the skin is smooth and thick with only a few pustules, the choanae are large and the inner ear is regressing (we have discussed the reduction of eyelashes in an aquatic environment), as are the teeth.

Most amphibians are able to exchange gases with water or air via their mucus-covered skin. For this cutaneous respiration to work, the surface of the skin is highly vascularised by the lymphatic and arterio-venous systems, and the cutaneous tissue must remain moist to allow oxygen to diffuse at a sufficiently high rate. Since the concentration of oxygen in water increases when the temperature is low and the flow rate is high, aquatic amphibians can, when these conditions are met, rely mainly on cutaneous respiration, as in the case of the Lake Tticaca frog[48] *Telmatobius culeus*, and the Euprocte des Pyrenees. In the open air, where oxygen is more concentrated, some small species can rely solely on cutaneous gas exchange to breathe, the most famous case being the salamanders of the **Plethodontidae** family, which have neither lungs nor gills. All amphibians have gills in their larval stage, and some aquatic salamanders retain them in their adult form. *Telmatobius culeus* is common in the Andes, in Bolivia, Ecuador, Chile, Argentina and Peru. This frog lives in Lake Titicaca at an altitude of 3,810 metres, and its skin is made up of numerous prominent skin folds. This strictly aquatic anuran breeds at the edge of the lake, preferring the bottom to feed on snails and aquatic insects. Adapted to a high altitude, it has a very low metabolic rate that requires less oxygen, and its breathing is mainly through the highly vascularised skin, which absorbs oxygen from the water through the large surface area of skin with folds that it stretches with brief jerking movements to increase absorption of the poorly oxygenated water, while unfolding very slowly. Ultimately, it will emerge from the water to breathe through its small, poorly vascularised lungs. Stressed, the skin secretes a sticky, viscous substance of low toxicity, sufficient to ward off predators. Its cutaneous respiratory sacs are highly vascularised on

---

48 Lake Titicaca is located in a tropical zone, at an altitude of 3,812 metres, with high exposure to solar UV radiation. It has a low oxygen concentration, with an average air temperature of around 11°7 to 12°7 and water temperature of 10°C. Oxygen varies with depth (thermoclines), for a pH of 8.34/8.93, the dissolved oxygen concentration is 6.02 to 7.29 g/l.

the back, sides and hind limbs. As it gains altitude, the skin thickens, with the epidermis composed of several cell layers that are larger than the dermis. In water, the skin is smooth with a few pustules, cutaneous respiration is predominant, the formation of cutaneous folds is accentuated by increasing the respiratory surface and there is a rich vascularisation of the epidermis with the presence of hematitic ampullae to capture oxygen from the rarefied air. In the adult, the skin is indeed rougher than in the larva, and we can agree that cutaneous respiration seems to act mainly at the level of the epidermal cells, extending to the connective dermis which is vascularised by the lymph. This skin adjustment variable is important for test modelling the evolution of Euproctes, because of the importance of cutaneous respiration and the local detoxification process that provides it with twofold protection, against desiccation in the open air and against predators. Although bacteria and pollution can rapidly affect its skin, it nevertheless assimilates ferric bacteria, which it finds beneficial.

Appendix 5

Life in an iron grip.

Euproctes observed in the larval stage in highly ferruginous ponds find their bodies and gills covered with fine, glowing silty particles loaded with iron. What are the possible consequences of this excessive concentration of iron? Can they take advantage of it for their respiration, in particular by feeding on ferrobacteria?

The springs and waterfalls of the Moudang are rich in metals in contact with outcropping metalliferous deposits. These metals, oxidised by acidophilic bacteria, produce acidified water loaded with fine particles of dissolved metals, including ferric iron (*Fe)*, which colours the Euproctes retention basins, springs and streams flowing into the Neste red. Bacteria such as ***Leptospirillum*** *ferriphilum* oxidise ferrous iron (Fe^{2+}) in pyrite into ferric iron (Battaglia-Brunet, 2010), making it one of the oldest low-free-energy redox systems involved in the biochemical processes of respiration. The redox potential of the Fe^{2+}/Fe^{3+} pair is 0.77 volts. I commented on the importance of this biochemical oscillator in the essay on low biological energies, and its presence in excess raised questions for me when observing Euproctes in their iron-saturated pools.

In most of the streams fed by springs with ferruginous water, in the absence of snails in the current, we have observed a remarkable development of aquatic plants spread out in long green filaments animated by an undulatory movement. In these highly oxygenated, well-lit waters, these plants are enriched with iron, a metal that carries electrons during the redox reactions that take place during photosynthesis, combining with enzymes to promote plant growth. Conversely, an iron deficiency would reduce growth and cause chlorosis. These aquatic plants are iron bio-accumulators. During photosynthesis, chlorophyll *a is* activated by light; at altitude and in high summer, absorption spectra of 400 nanometres for ultra-violet and 700 nanometres for infrared accelerate the reduction of $NADP+$ at the brightest times of day. In photosystem *I,* the electron released by the chlorophyll *has* passed into an iron-containing protein, ferrodoxin, which is reduced and rë-oxidised in a flavoprotein, transferring the electron to ***NADP'***, which gains a hydrogen atom *H*. Consecutively, $NADPH_{49}$ is used to reduce CO_2 to carbohydrates during the dark phase of photosynthesis. In these streams, the iron content would reach an optimal value for the development of these aquatic plants in correlation with the high summer luminosity at altitude, benefiting from the high oxygenation induced by the turbulent flow of the water.

49 Nicotinamide Adenine Dinucleotide Phosphate Hydrogen, NADPH-oxidase is a membrane enzyme complex.

However, for aquatic animals, this presence of iron would exceed toxicity thresholds, which are certainly higher during the summer, in the calm waters of Euproctes ponds, where larvae with gills and adults in cutaneous and bucco-pharyngeal respiration mode are present.

Compare the Pyrenean Euprotect with the cave-dwelling *Proteus anguinus,* Laurenti, 1768, a primitive cave-dwelling Urodele amphibian that remains in the larval stage, equipped with gills, and lives exclusively in the water. Its skin is white to slightly pink, covered with *"a hyaline mucus that becomes opaque in the event of stress"*. The photosensitive skin pigments from dark grey to black in the light. Found in south-east Italy, from Slovenia to Croatia and Bosnia-Herzegovina, it lives in deep, dark cave waters at a temperature of between 5°C and 14°C, with a basic pH of 7.4 to 8.7 and a dissolved oxygen level of 7 to 12 mg/L. Among its food, it absorbs an anaerobic and autotrophic ferrobacterium *Perabacterium* spelaei (Victor Caumartin, 1957), which fixes nitrogen in the air. This bacterium obtains its energy and carbon from the decomposition of iron carbonate when ferrous oxide is released, and from its decomposition into ferric oxide when dioxygen is released from water. This ferrobacteria, included in the adsorbent silty clay, supplies trace elements, vitamins and oxygen. The Protea is a very long-lived young organism that engages in autophagy, so assimilates these bacterial by-products in order to survive. Note the photosensitivity of the mucus and skin as a respiratory organ complementary to the gills, as well as the gain of oxygen linked to the iron carbonate, in low-oxygen, basic water. Compared with the Euproctus, a facultative cave-dweller at medium altitude, it consumes half as much oxygen at minimal activity, and therefore has a low-energy metabolism. Segmentation of the egg, which has an initial diameter of around 5 mm, lasts 15 minutes at 11°6 C until the blastula stage. Its breathing is essentially gill-based, the lungs are not functional, and outside the water, cutaneous respiration predominates.

Neoteny (Kolmann, 1884) in the Protea and Axolotl (*Ambystoma mexicanus*) allows reproduction in the larval stage due to a thyroid insensitivity of ancient origin. The metamorphosis of the Urodeles is thought to correspond to an adaptation of the amphibians that enabled them to make the transition from aquatic life to terrestrial life, which appeared around the middle of the Tertiary period, and therefore to a temperature- and light-dependent activation of thyroid activity in response to stressful situations, via iodine, which is genetically inactivated in species subject to permanent neoteny.

At a normal pH of 7, iron precipitates at concentrations of between 4.5 and 13.5 mg/l, and does not precipitate between 0.5 and 1.5 mg/l. Iron at 50 mg/l is highly toxic to larvae, causing delayed growth and death from a form of hemosiderosis. Ferrous salts are unstable in air, and **"oxygenated iron oxide absorbs protons"** to give ferrous oxide and water, which contributes to tissue hydration and helps produce mucus in skin tissue by reducing the acidity of the environment, which modulates the activity of enzymes in the body's cells. The iron atom has twenty-six electrons, and the ferrous ion in aqueous solution acquires electrons from the oxygen atoms in the water or air saturated with water to form an active complex ion. Blood and the cellular milieu are aqueous buffer media which absorb iron; it becomes active in the presence of oxygen, by mobilising the translocation of protons and the transfer of electrons during respiration.

To give an idea of the magnitude of iron toxicity, in humans whose tolerance to iron is only 100 micrograms per litre, iron overload causes hemosiderosis, which affects the colour of the skin and causes irreversible anatomical and functional lesions. The table of pathologies is briefly summarised below for the known target organs, at the level of :

- In the pituitary gland, it causes stunted growth.
- Parathyroi'de, a hyperparathyroi'die.
- Heart disease, heart failure and other disorders.
- Liver, Cytolysis and Cirrhosis.
- Pancreas, diabetes.
- Gonads, hypogonadism and infertility.
- Mutations, the probable causes of which we discuss below.

Although iron is not very soluble in water, as we have just seen, its ferrous salt form is more soluble than that of ferric salts. It is absorbed in a colloidal state by the skin and the buccal-pharyngeal route of the Euproctus. The silty substrates are associated with minerals, vegetation, organic waste and ferrobacteria that the Euproctus larvae consume in varying proportions depending on the season in the ponds where they live. These natural, shallow, iron-laden silty pools are home to ferrobacteria that absorb and oxidise the iron, releasing oxygen. Does the presence of these iron bacteria play a role in limiting the pathologies caused by excess iron? Do they play a part in the metabolism and cutaneous respiration of the Euproctes that feed on them in the larval stage in iron-rich ponds?

Iron is present in metabolic functions, in the form of ferric oxide and ferrous oxide. This oscillating biochemical couple is involved in oxidation-reduction reactions with low free energy. At the cellular level, iron catalyses enzymes that contribute to electron transport, oxygen activation and nucleotide and *DNA* synthesis, depending on the free energy supplied by *ATP* in conjunction with proton translocation in the mitochondria. Iron oxide Fe^{3+} is not very soluble at physiological pH, and proteins are involved in its solubilisation and transport within the cell, with intracellular storage in the form of ferritin. It is transported extra-cellularly using transferrin, and the cell's plasma membrane contains transferrin receptors that capture ferrotransferrin. After endocytosis, iron acts in the cellular environment at the request of the cell and the organism, under the control of the nervous and hormonal systems, in response to attacks from the external environment. It acts in the cell in association with ferritin, or in part, in a low molecular weight complex. This complex regulates the metabolism of this metal by adjusting the number of these receptors, either when there is a deficiency or an excess of iron, in our case of study.

In the inner membrane of the mitochondria, it is present in the cytochromes, as a mobile extra-membrane electron transporter between complex III (cytochrome *bc*) and complex IV (cytochrome oxidase). Rejection of cytochrome *c is* involved in apoptosis, programmed cell death, as cytochrome *c* is released into the cytoplasm. Cytochrome is a ferroporphyrin coenzyme involved in the oxidation-reduction process, catalysed by a reversible reaction from ferric oxide to ferrous oxide.

Excess iron is an oxidative stress factor that could promote the production of free radicals, the reactive oxygen species (ROS).

In the presence of oxygen, iron in its ferric ion form has the particularity of transforming into ferric hydroxide, which is particularly insoluble and would therefore be a limiting factor for development. However, micro-bacteria possess sidërophores to capture ferric hydroxide and transport it into the cells, and Euproctes larvae confined in their iron-saturated tanks consume these bacteria, thereby limiting the toxicity of iron on growth.

What is the role of ventral (orange to red) and dorsal (yellow and black) coloured pigments that are photosensitive to light?

While adult Euproctes live in highly oxygenated waters, larvae equipped with gills move around in small pools with less flowing water. Skin respiration provides oxygen by diffusion into the tissues via the cells of the epidermis and dermis. Respiratory pigments such as ferritin and siderophilin are involved in the storage and transfer of oxygen into the blood, where iron-containing haemoglobin, specific to each species, varies with age. Its structure is in a normal or pathological state depending on the iron content and the variation in physico-chemical factors in the environment (pH, temperature, chemical complexes associated with iron, natural or industrial pollution). Hemoglobin combines with oxygen in a reversible manner to give oxyhemoglobin in relation to the amount of iron. This oxygenation of ferrous oxide depends on atmospheric pressure and the concentration of oxygen in the air, which varies with altitude, depth and turbidity of the water. The oxygen present in the water diffuses through the epidermis from the skin cells to the connective tissue cells of the dermis, where the lymph is in contact with the capillaries of the general bloodstream. In the blood, oxygen pressure is lower than in the skin cells, but carbon dioxide pressure rises, as does the concentration of H ions, releasing oxygen into the tissue cells and organs. As a result, carbon dioxide is evacuated from the connective tissue by the skin cells through direct detoxification by the epidermal cells and excretory glands, and secondarily by the general circulation in the connective tissue lining the oral-pharyngeal breathing cavity, or occasionally by the residual lungs required when going out of the water. Oxygenation and direct detoxination localized in the skin cells are thus supplemented by these last two forms of respiration, which are more adapted to life on land, in particular pulmonic respiration, which for many species will acquire an increasingly effective ability to capture large quantities of air, increasing oxygen pressure and above all evacuating carbon dioxide, which helps to maintain the potential of $H+$ *ions at* an acceptable level of acidity for the organism.

We have just asked questions about the impact of excess iron toxicity on metabolism, but we must also ask about its possible mutagenic effects. Could this sensitivity have indirect effects on Euproctes larvae?

In reaction to variations in the internal environment and fluctuations in the environment on the nervous system, hormones and the immune system condition the responses of the cells to oxidative toxic stress caused by iron, which is very active in the body and at cellular level. As we have just seen, iron is involved in localized cutaneous cellular respiration, through differential coupling of the lymphatic system to the general blood circulation of the Euproctus, acting on enzymes and participating in oxidative mëtabolism. As iron is present in ribonuclease reductase, which catalyses the reduction of ribonucleotides to dësoxyrobonuclëides, the building blocks of DNA, this raises another question about its gënëtic toxicity?

In saturated basins, residues of iron oxide, *Fe,* which are not very soluble, become soluble through the action of specific proteins and possibly the bacteria mentioned above. As a result, they are transported out of the cells by transferrin and stored in the liver and spleen cells in the form of ferritin. Receptors in the plasma membrane capture the ferrotransferrin and, by endocytosis, admit the iron into the cell to satisfy its function, thus meeting the body's iron requirements. In the cell, if the iron remains bound to ferritin, a fairly minute part of low molecular weight, in the form of a non-negligible complex, would act as a regulator of the metabolism of this metal.

In the event of iron overload, the membrane receptors activated by the excess iron would restrict ferritin molecules, but to what limit in the Euproctus?

We have just seen that iron is necessary for metabolism and respiration, and is involved in DNA synthesis, contributing to cell proliferation. Research is tending to show that specific antibodies block transferrin membrane receptors, and are also capable of stopping or at least reducing tumour proliferation, depending on the species. According to these studies, an increase in the level of iron in cells is an aggravating factor in the risk of cancer and chromosomal abnormalities. Toxicity tests with iron at 13.5 mg/l give an abnormal mortality rate, and ecotoxicity occurs at high densities and induces oxidative stress (Fabrice Godet, 1993). Iron is thought to act on the mitotic spindle, causing DNA breaks. Iron also catalyses the production of highly toxic free radicals (hydroxyl OH, superoxide anion O^{2-}), highly reactive free radicals have a very short lifespan, and if they are electrically neutral, the single electron causes incident paramagnetism likely to modify the structure of surrounding atoms.

These free radicals are produced by thermal, photochemical or radiochemical activation, as well as by biochemical reaction with slow electrons, for example in a homolytic break-up of a chemical species, the two resulting products being free radicals, generators of chain reactions. Glutathione is known to cause oxidative stress by producing **OH** and o_2, which are highly toxic to cells. It plays an important role in detoxification and in inducing oxidative stress. Oxygen-reactive species are thought to be involved in mutagenesis and cancerogenesis. Iron is thought to indirectly induce alterations in genetic material and retarded growth.

To sum up, in addition to its impact on growth and serious pathological effects, the induction of oxidative stress by excess iron cannot be neglected. Without being directly toxic to the genes, iron, by producing highly reactive oxygen radicals, causes disturbances to the genetic material, such as DNA breaks, leading indirectly to mutagenic effects, as well as chromosomal aberrations that may be deleterious or selectively favourable to the remarkably stabilising evolution of Euproctes and Urodeles.

Bibliography

- Abeloos Marcel, *Les Metamorphoses*, Collection Armand Collin, 1956.

-Abi Nahed Roland, *Phospholipases A2 during fecundation and preimplantation embryonic development: molecular mechanism of action and therapeutic development*, These docteur biologie du dëveloppement- oncogenese, universite de Grenoble, 2015.

- Arneodo Alain, Argoul Frangoise, Bacry Emmanuel, Elezgaray Juan, Muzy Jean-Francois, *Ondelettes, multifractales et turbulences*, Diderot, 1995.

-Battaglia-Brunet, *Microflore bacterienne des milieux riches en metaux et metalloides*, Universite de Provence-Aix-Marseille I, 2010.

- Bautz Alain, *Role presume de la gravite dans la symetrisation de l'embryon des amphibiens, reponse apportee par l'experimentation en biologie dans l'espace*, Bulletin de l'Academie Lorraine des Sciences, 2002.

- Beetschen Jean Claude, *l'Embryologie*, que sais je, puf, 1979.

- Berg Alonso Laetitia, *Mitochondrial respiratory chain deficiencies with mitochondrial DNA instability: identification of novel genes and mechanisms*, Universite de Nice Sophia Antipolis, 2016.

- Brackett Frederick S., *The present state of physics*, American association for the advancement of Science, 1954.

- Caraguel Flavien, *Proliferation during regeneration of the lepidogenic bilobed form of the skin fin in zebrafish*, These Universite de Grenoble, 2006.

- Chaline Jean, Nottale Laurent, Grou Pierre, *Des fleurs pour Schrodinger*, ellipses, 2009.

- Cohen-Tannoudji Gilles and Spiro Michel, *Le boson et le chapeau mexicain*, Gallimard, 2013.

- Cottet-Roussselle Cecile, *Mesure par microscopie confocale du metabolisme mitochondrial et du niveau energetique cellulaire au cours des episode de carences en substrats et/ou en oxygene*, These de Doctorat de Biologie cellulaire, Moleculaire et Sciences de la Sante, Ecole Pratique des Hautes etudes et Universite de Grenoble Alpes, 2013.

- Coulon Antoine, *Stochasticity of gene expression and transcriptional regulation*, Institut National des Sciences Appliquees, 2010.

- Cox Brian and Forshaw Jeff, *The Quantum Universe*, Dunod, 2013.

- Cunchillos Chomin, *Les voies de l'emergence*, Belin, 2014.

- Dehaut E.G., *Les venins des batraciens, etude de zoologie medicale*, G. Steinheil editeur, 1969.

- De Nadai Celine, Chiri Sandrine, Brigitte Ciapa, *Les mecanismes de l'activation ovocytaire*, Synthese medecine sciences, 1999.

- Dendaletche Claude, *L'Homme et la Nature dans les Pyrenees,* Berger-Levrault, 1982.

- David Patrice and Samadi Sarha, *The theory of revolution*, Flammarion, 2006.

- Despax, L *Euprocte des Pyrenees*, bulletin de la societe d'histoire naturelle de Toulouse.

- Despax R., *Note sur la vascularisation de la peau chez l'Euprocte des Pyrenees*, bulletin de la sociëtë zoologique de France, 1914.

- Devillers C. and Chaline J., *La theorie de revolution*, Dunod, 1989.

- Feynman Richard, *Lumiere et matiere,* Intereditions, 1987.

-Fiszman Pierre-Louis, *Les mecanismes physiologiques responsables de l'Homochromie variable dans le regne animal*, Thëse pour le doctorat vëtërinaire, faculte de Mëdecine de Creteil, 2017.

- Fleury Vincent, *La chose humaine*, Vuibert, 2009.

- Godet Fabrice, *etude de l'ecotoxicite d'effluents complexes*, Université Paul Verlaine, M4etz, 1993.

- Guillaume Olivier, *Role de la communication interspecifique dans la segregation spatiale chez les Urodeles sympatriques Calotriton asper asper et Salamandra salamandra*, Bulletin de la societe herpetologique de France, n°18, 2006.

- Guille-Escuret Georges, *Les societes et leurs natures*, Armand Colin, 1989.

- Guille-Escuret Georges, *Le decalage humain,* published by Kime, 1994.

- Haldane JBS, Bernal J.D.,Pirie N.W., Pringle J.W.S., *A Discussion on the Origin of Life*, Rationalist Union Publications, 1955.

- Julienne Cloe Minsy, *Alteration du metabolisme energetique mitochondrial lors de la cachexie cancereuse*, Universite Francois Rabelais de Tours, 2012.

- Kupiec Jean-Jacques, *L'origine des individus,* Fayard, 2008.

- Le Garff Bernard, *Biology and ecology of amphibians*, 2020.

- Laffitte M. and Rouquerol, *La reaction chimique*, Masson, 1990.

- Lamotte Maxime, *Theorie actuelle de revolution*, Hachette, 1994.

- Le Douarin Nicole, *Dictionnaire amoureux de la vie*, Plon, 2017.

- Lemieux Helene, *Effets de la température sur le metabolisme mitochondrial cardiaque*, Universite du Quebec a Rimouski, 2007.

- Majupuria T.C. AND Rohit Kumar, *WILDLIFE and protect aeras of Nepal*, DEVI,2006.

- Maarinez Rica Juan-Pablo and Clergue-Cazeau Monique, *Donnees nouvelles sur la repartition geographique de l'espece Euproctus asper (Duges)*, Bulletin de la societe d'histoire naturelle de Toulouse, 1977.

- Mayr Enrst, *Populations, especes et evolution*, Hermann, 1974.

- Ndiaye Dieynaba, *Role des mitochondries dans la régulation des oscillations de calcium des hepatocytes: approches experimentale et computationnelle*, Universites Paris-Sud and Libre de Bruxelles, 2013.

- Nonnenmacher Stephane, *Quelques aspects du chaos quantique, Memoire physique et mathematiques*, Universite Paris-Sud, Faculte des sciences d'Orsay, 2009.

- Pierre Virginie, *L'acrosome du spermatozoide, de sa biogenese a son rôle physiologique*, These docteur universite de Grenoble, 2013.

- Roland J.-C and Szollosi A. et D., *Atlas de biologie cellulaire*, Masson, 1986.

- Sacchi C.F. and P. Testard, *ecologie animale*, Doin editeur, 1971.

- Schlegel Peter A ., Briegleb Wolfgang, Bulog Boris, Steinfartz Sebastian, *Revue et nouvelles donnees sur la sensibilite a la lumiere et orientation non visuelle chez Proteus anguinus, Calotriton asper et Desmognathus ochrophaeus (Amphibiens urodeles hypoges)*, Bulletin de la societe herpetologique de France, n°18, 2006.

- Silvestre Marc, *L 'etevage des Urodeles: etude de cinq especes menacées*, These pour le doctorat veterinaire, Faculte de medecine de Creteil, 2001.

- Tort Patrick, *Darwinisme et societe*, puf, 1992.

- Weil j-H, *biochimie generale*, Masson, 1983.

Illustrations

- Cover page: An Euproctus breathes at the surface of the water.
- Page of thanks : Embleme de l'auteur " La dame a l'ancre ".
- Page 14, Preface: Gun muzzle tape from CIN Querqueville.
- Foreword page: Ancre de Marine.
- 1. Hautes-Pyrenees, the Moudang valley.
- 2. Library, Charles Darwin Institute International.
- 3. The Pyrenees, up the Vallee d'Aure.
- 4. Salamander carved on reindeer antler, Laugerie-Basse rock shelter, Les Eyzies de Tayac, Dordogne.
- 5. A flock of sheep grazing at altitude.
- 6. Extract from the Cassini map of Thury n°76/21F/Bourgoin, vallee d'Aure, from the archives of Pau, Bearn, Pyrenees.
- 7. Source ferrugineuse de la Reine.
- 8. On the way to the Moudang valley.
- 9. Barns in the Moudang valley.
- 10. The Neste and the Moudang barns.
- 11. A flock of sheep around the barns.
- 12. Waterfall that feeds the Euproctes and Neste basins below.
- 13 and 14. Miraculous spring of the Queen and deposed virgins.
- 15 and 16. Development of aquatic plants in excess of iron at the source of the Queen.
- 17. Cows grazing at altitude.
- 18. Marmot on the lookout.
- 20. Griffon vulture in flight.
- 21. Isards, Rupicapra pyrenaica, on the heights.
- 22. Pair of marmots and lookout on a rock.
- 23. A high-altitude farm in Bhutan.
- 24. Red frog and Pyrenean snake.
- 25. Diagram of the evolution of the Urodeles branch.
- 26. Euproctus, Calotriton asper asper, in a ferriginous basin.
- 27. Axolotl, Ambystoma mexicanum.
- 28. Euproct larva with gills.
- 29. Euproctes basin, below a waterfall.
- 30. Details of epidermal protuberances.
- 31. Schema of Raymond Dexpax, epidermis and dermis of the skin of the Euproctus.
- 32. Cross-section of the skin of an Urodele.
- 33. Adult euproct, dorsal line of xantophores.
- 34. Summary diagram of the skin.
- 35. Lymphatic burrows (L) of Rana escalata.
- 36. Terrestrial salamander, Salamandra salamandra.
- 37. and 38. Common toad, Buf bufo and toad half eaten by a predator.
- 39 and 40. Red frog and skinned frog.
- 41. Mucus on the skin of a red frog (Mont Lozere, 1,400 metres).
- 42. Mating of Euproctes.
- 43. Cluster of red frog eggs (Mont Lozere, 1,400 metres).

-     44. fёcondёs frog eggs (Mont Lozere).
-     45. Development of red frog eggs.
-     46. Larvae with red frog gills.
-     47. Experimental calcium signal diagram.
-     48. Segmentation phases of the Axolotl reuf.
-     49. frog eggs and larvae (Mont Lozere).
-     50. Test model.
-     51. Table and diagram of the divergent segmentation of micromeres
at each pole.
-     52. Sea anemone fixed to the seabed, a good bait for tides.
-     53. Metamorphosis of the red-legged frog on the river Ceze (400 metres).
-     54. Tres gros tetards.
-     55 and 56. Larvae with frog and salamander gills.
-     57. Pyrenean snail larva with gills.
-     58 and 59. Fossils, Tetrapod skull and Ichtyostea.
-     60. Regeneration of a starfish.
-     61. In the Euproctus, a limb that has been severed incidentally regenerates itself.
-     62. synthetic dna 1.
-     63. Synthetic DNA 2.
-     64. Synthetic DNA 3.
-     65. Additive solution.
-     66. A minimum of energy.
-     67. ffiil of Euprocte.
-     68. In a rocky basin, Euproctes hunt, raised from the depths.
-     69. Euproctes hunt insects in the current.
-     70, 71 and 72. Environment of the Pyrenean skink.
-     73. Simplified approach to the phylogeny of lissamphibians, divergence of Anurans
and Urodeles.
-     74. Comparative table of parameters.
-     75. The terrestrial solution.
-     76. Simulation of the reversionary effect.
-     77. Lissamphibians tree diagram.
-     Back cover, author Claude Rouquette.

THE PYRENEAN LOUSEWORT
Claude ROUQUETTE

While exploring the Hautes-Pyrenees (France), naturalist Claude ROUQUETTE came across the Pyrenean skink Calotriton asper asper near Pont-de-Moudang, just as he was beginning to develop his research method on the complex processes involved in biological evolution and the transformations of civilisation.

From question to question, he takes us through the skin respiration, development and metamorphosis of the Euproctus, examining its ability to regenerate an organ, and commenting on the beginnings of their sociability, which he observed in a pond where these mythical salamanders hunt.

After a career as an officer in the French Navy, Claude ROUQUETTE, a naval historian and naturalist, has given concrete form to his long-term research into complex evolutionary processes by writing a naturalist suite in four volumes, supplemented by three essays. It promotes long-term scientific exploration at sea and on land.

I want morebooks!

Buy your books fast and straightforward online - at one of world's fastest growing online book stores! Environmentally sound due to Print-on-Demand technologies.

Buy your books online at
www.morebooks.shop

Kaufen Sie Ihre Bücher schnell und unkompliziert online – auf einer der am schnellsten wachsenden Buchhandelsplattformen weltweit! Dank Print-On-Demand umwelt- und ressourcenschonend produzi ert.

Bücher schneller online kaufen
www.morebooks.shop

Printed by Books on Demand GmbH, Norderstedt / Germany